Student Lab Ma

for
Argument-Driven Inquiry
in
EARTH AND SPACE SCIENCE

LAB INVESTIGATIONS
for GRADES 6–10

Student Lab Manual

for
Argument-Driven Inquiry
in
EARTH AND SPACE SCIENCE

LAB INVESTIGATIONS
for GRADES 6–10

Victor Sampson, Ashley Murphy, Kemper Lipscomb, and Todd L. Hutner

NSTApress

National Science Teachers Association
Arlington, Virginia

National Science Teachers Association

Claire Reinburg, Director
Rachel Ledbetter, Managing Editor
Deborah Siegel, Associate Editor
Andrea Silen, Associate Editor
Donna Yudkin, Book Acquisitions Manager

ART AND DESIGN
Will Thomas Jr., Director

PRINTING AND PRODUCTION
Catherine Lorrain, Director

NATIONAL SCIENCE TEACHERS ASSOCIATION
David L. Evans, Executive Director

1840 Wilson Blvd., Arlington, VA 22201
www.nsta.org/store
For customer service inquiries, please call 800-277-5300.

NSTA is committed to publishing material that promotes the best in inquiry-based science education. However, conditions of actual use may vary, and the safety procedures and practices described in this book are intended to serve only as a guide. Additional precautionary measures may be required. NSTA and the authors do not warrant or represent that the procedures and practices in this book meet any safety code or standard of federal, state, or local regulations. NSTA and the authors disclaim any liability for personal injury or damage to property arising out of or relating to the use of this book, including any of the recommendations, instructions, or materials contained therein.

Cataloging-in-Publication Data are available from the Library of Congress.
LCCN: 2017055013
ISBN: 978-1-68140-598-8
e-ISBN: 978-1-68140-599-5

CONTENTS

SECTION 1

Introduction and Lab Safety

SECTION 2

Space Systems

INTRODUCTION LABS

Lab 1. Moon Phases: Why Does the Appearance of the Moon Change Over Time in a Predictable Pattern?

Lab 2. Seasons: What Causes the Differences in Average Temperature and the Changes in Day Length That We Associate With the Change in Seasons on Earth?

Lab 3. Gravity and Orbits: How Does Changing the Mass and Velocity of a Satellite and the Mass of the Object That It Revolves Around Affect the Nature of the Satellite's Orbit?

APPLICATION LAB

Lab 4. Habitable Worlds: Where Should NASA Send a Probe to Look for Life?

SECTION 3
History of Earth

INTRODUCTION LABS

APPLICATION LAB

SECTION 4
Earth's Systems

INTRODUCTION LABS

APPLICATION LABS

SECTION 5

Weather and Climate

INTRODUCTION LABS

APPLICATION LAB

Lab 19. Differences in Regional Climate: Why Do Two Cities Located at the Same Latitude and Near a Body of Water Have Such Different Climates?

SECTION 6

Human Impact

INTRODUCTION LABS

Lab 20. Predicting Hurricane Strength: How Can Someone Predict Changes in Hurricane Wind Speed Over Time?

Lab 21. Forecasting Extreme Weather: When and Under What Atmospheric Conditions Are Tornadoes Likely to Develop in the Oklahoma City Area?

APPLICATION LABS

Lab 22. Minimizing Carbon Emissions: What Type of Greenhouse Gas Emission Reduction Policy Will Different Regions of the World Need to Adopt to Prevent the Average Global Surface Temperature on Earth From Increasing by 2°C Between Now and the Year 2100?

Lab 23. Human Use of Natural Resources: Which Combination of Water Use Policies Will Ensure That the Phoenix Metropolitan Area Water Supply Is Sustainable?

ABOUT THE AUTHORS

Victor Sampson is an associate professor of STEM (science, technology, engineering, and mathematics) education and the director of the Center for STEM Education at The University of Texas at Austin (UT-Austin). He received a BA in zoology from the University of Washington, an MIT from Seattle University, and a PhD in curriculum and instruction with a specialization in science education from Arizona State University. Victor also taught high school biology and chemistry for nine years. He specializes in argumentation in science education, teacher learning, and assessment. To learn more about his work in science education, go to *www.vicsampson.com.*

Ashley Murphy attended Florida State University (FSU) and earned a BS with dual majors in biology and secondary science education. Ashley spent some time as a middle school biology and science teacher before entering graduate school at UT-Austin, where she is currently working toward a PhD in STEM education. Her research interests include argumentation in middle and elementary classrooms. As an educator, she frequently employed argumentation as a means to enhance student understanding of concepts and science literacy.

Kemper Lipscomb is a doctoral student studying STEM education at UT-Austin. She received a BS in biology and secondary education from FSU and taught high school biology, anatomy, and physiology for three years. She specializes in argumentation in science education and computational thinking.

Todd L. Hutner is the assistant director for teacher education and center development for the Center of STEM Education at UT-Austin. He received a BS and an MS in science education from FSU and a PhD in curriculum and instruction from UT-Austin. Todd's classroom teaching experience includes teaching chemistry, physics, and Advanced Placement (AP) physics in Texas and earth science and astronomy in Florida. His current research focuses on the impact of both teacher education and education policy on the teaching practice of secondary science teachers.

SECTION 1
Introduction and Lab Safety

INTRODUCTION

Science is much more than a collection of facts or ideas that describe what we know about how the world works and why it works that way. Science is also a set of crosscutting concepts and practices that people can use to develop and refine new explanations for, or descriptions of, the natural world. These core ideas, crosscutting concepts, and scientific practices are important for you to learn. When you understand these core ideas, crosscutting concepts, and scientific practices, it is easier to appreciate the beauty and wonder of science, to engage in public discussions about science, and to critique the merits of scientific findings that are presented through the popular media. You will also have the knowledge and skills that you will need to continue to learn about science outside school or to enter a career in science, engineering, or technology once you learn the core ideas, crosscutting concepts, and scientific practices.

The core ideas of science include the theories, laws, and models that scientists use to explain natural phenomena or to predict the results of new investigations. The crosscutting concepts of science, in contrast, are themes that have value in every discipline of science as a way to help understand a natural phenomenon. These crosscutting concepts can be used as an organizational framework for connecting knowledge from the various fields of science into a coherent and scientifically based view of the world and to help us consider what is important to think about or look for during an investigation. Finally, scientific practices are used to develop and refine new ideas about the world. Although some practices differ from one field of science to another, all fields share a set of common practices. The practices include such things as asking and answering questions, planning and carrying out investigations, analyzing and interpreting data, and obtaining, evaluating, and communicating information. One of the most important scientific practices is arguing from evidence. Arguing from evidence, or the process of proposing, supporting, challenging, and refining claims based on evidence, is important because scientists need to be able to examine, review, and evaluate their own knowledge and ideas and critique those of others. Scientists also argue from evidence when they need to appraise the quality of data, develop and refine models, develop new testable questions from those models, and suggest ways to refine or modify existing theories, laws, and models.

It is important to always remember that science is a social activity, not an individual one. Science is a social activity because many different scientists contribute to the development of new scientific knowledge. As scientists carry out their research, they frequently talk with their colleagues, both formally and informally. They exchange e-mails, engage in discussions at conferences, share research techniques and analytical procedures, and present new ideas by writing articles in journals or chapters in books. They also critique the ideas and methods used by other scientists through a formal peer-review process before they can be published in journals or books. In short, scientists are members of a community, and the members of that community work together to build, develop, test, critique, and refine ideas. The ways scientists talk, write, think, and interact with each other reflect common ideas about what counts as quality and shared standards for how new ideas should be developed, shared, evaluated, and refined. These ways of talking, writing, thinking, and interacting make science different from other ways of knowing. The core ideas, crosscutting concepts, and scientific practices are important within the scientific community because most, if not all, of the members of that community find them to be a useful way to develop and refine new explanations for, or descriptions of, the natural world.

The laboratory investigations that are included in this book are designed to help you learn the core ideas, crosscutting concepts, and scientific practices that are important in Earth and space science. During each investigation, you will have an opportunity to use a core idea and several crosscutting concepts and scientific practices in order to understand a natural phenomenon or to solve a problem. Your teacher will introduce each investigation by giving you a task to accomplish and a guiding question to answer. You will then work as part of a team to plan and carry out an investigation to collect the data that you need to answer that question. From there, your team will develop an initial argument that includes a claim, evidence in support of your claim, and a justification of your evidence. The claim will be your answer to the guiding question, the evidence will include your analysis of the data that you collected and an interpretation of your analysis, and the justification will explain why your evidence is important. Next, you will have an opportunity to share your argument with your classmates and critique their arguments, much like professional scientists do. You will then revise your initial argument based on their feedback. Finally, you will be asked to write an investigation report on your own to share what you learned. These reports will go through a double-blind peer review so you can improve it before you submit to your teacher for a grade. As you complete more and more investigations in this lab manual, you will not only learn the core ideas associated with each investigation but also get better at using crosscutting concepts and scientific practices to understand the natural world.

SAFETY IN THE SCIENCE CLASSROOM, LABORATORY, AND FIELD SITES

Note to science teachers and supervisors/ administrators:

The following safety acknowledgment form is for your use in the classroom and should be given to students at the beginning of the school year to help them understand their role in ensuring a safer and productive science experience.

Science is a process of discovering and exploring the natural world. Exploration occurs in the classroom/laboratory or in the field. As part of your science class, you will be doing many activities and investigations that will involve the use of various materials, equipment, and chemicals. Safety in the science classroom, laboratory, or field sites is the FIRST PRIORITY for students, instructors, and parents. To ensure safer classroom/laboratory/field experiences, the following **Science Rules and Regulations** have been developed for the protection and safety of all. Your instructor will provide additional rules for specific situations or settings. The rules and regulations must be followed at all times. After you have reviewed them with your instructor, read and review the rules and regulations with your parent/guardian. Their signature and your signature on the safety acknowledgment form are required before you will be permitted to participate in any activities or investigations. Your signature indicates that you have read these rules and regulations, understand them, and agree to follow them at all times while working in the classroom/laboratory or in the field.

Safety Standards of Student Conduct in the Classroom, Laboratory, and in the Field

1. Conduct yourself in a responsible manner at all times. Frivolous activities, mischievous behavior, throwing items, and conducting pranks are prohibited.

2. Lab and safety information and procedures must be read ahead of time. All verbal and written instructions shall be followed in carrying out the activity or investigation.

3. Eating, drinking, gum chewing, applying cosmetics, manipulating contact lenses, and other unsafe activities are not permitted in the laboratory.

4. Working in the laboratory without the instructor present is prohibited.

5. Unauthorized activities or investigations are prohibited. Unsupervised work is not permitted.

6. Entering preparation or chemical storage areas is prohibited at all times.

7. Removing chemicals or equipment from the classroom or laboratory is prohibited unless authorized by the instructor.

Personal Safety

8. Sanitized indirectly vented chemical splash goggles or safety glasses as appropriate (meeting the ANSI Z87.1 standard) shall be worn during activities or demonstrations in the classroom, laboratory, or field, including pre-laboratory work and clean-up, unless the instructor specifically states that the activity or demonstration does not require the use of eye protection.

9. When an activity requires the use of laboratory aprons, the apron shall be appropriate to the size of the student and the hazard associated with the activity or investigation. The apron shall remain tied throughout the activity or investigation.

10. All accidents, chemical spills, and injuries must be reported immediately to the instructor, no matter how trivial they may seem at the time. Follow your instructor's directions for immediate treatment.

11. Dress appropriately for laboratory work by protecting your body with clothing and shoes. This means that you should use hair ties to tie back long hair and tuck into the collar. Do not wear loose or baggy clothing or dangling jewelry on laboratory days. Acrylic nails are also a safety hazard near heat sources and should not be used. Sandals or open-toed shoes are not to be worn during any lab activities. Refer to pre-lab instructions. If in doubt, ask!

12. Know the location of all safety equipment in the room. This includes eye wash stations, the deluge shower, fire extinguishers, the fume hood, and the safety blanket. Know the location of emergency master electric and gas shut offs and exits.

13. Certain classrooms may have living organisms including plants in aquaria or other containers. Students must not handle organisms without specific instructor authorization. Wash your hands with soap and water after handling organisms and plants.

14. When an activity or investigation requires the use of laboratory gloves for hand protection, the gloves shall be appropriate for the hazard and worn throughout the activity.

Specific Safety Precautions Involving Chemicals and Lab Equipment

15. Avoid inhaling fumes that may be generated during an activity or investigation.

16. Never fill pipettes by mouth suction. Always use the suction bulbs or pumps.

17. Do not force glass tubing into rubber stoppers. Use glycerin as a lubricant and hold the tubing with a towel as you ease the glass into the stopper.

18. Proper procedures shall be followed when using any heating or flame-producing device, especially gas burners. Never leave a flame unattended.

19. Remember that hot glass looks the same as cold glass. After heating, glass remains hot for a very long time. Determine if an object is hot by placing your hand close to the object but do not touch it.

20. Should a fire drill, lockdown, or other emergency occur during an investigation or activity, make sure you turn off all gas burners and electrical equipment. During an evacuation emergency, exit the room as directed. During a lockdown, move out of the line of sight from doors and windows if possible or as directed.

21. Always read the reagent bottle labels twice before you use the reagent. Be certain the chemical you use is the correct one.

22. Replace the top on any reagent bottle as soon as you have finished using it and return the reagent to the designated location.

23. Do not return unused chemicals to the reagent container. Follow the instructor's directions for the storage or disposal of these materials.

Standards for Maintaining a Safer Laboratory Environment

24. Backpacks and books are to remain in an area designated by the instructor and shall not be brought into the laboratory area.

25. Never sit on laboratory tables.

26. Work areas should be kept clean and neat at all times. Work surfaces are to be cleaned at the end of each laboratory or activity.

27. Solid chemicals, metals, matches, filter papers, broken glass, and other materials designated by the instructor are to be deposited in the proper waste containers, not in the sink. Follow your instructor's directions for disposal of waste.

28. Sinks are to be used for the disposal of water and those solutions designated by the instructor. Other solutions must be placed in the designated waste disposal containers.

29. Glassware is to be washed with hot, soapy water and scrubbed with the appropriate type and sized brush, rinsed, dried, and returned to its original location.

30. Goggles are to be worn during the activity or investigation, clean up, and through hand washing.

31. Safety Data Sheets (SDSs) contain critical information about hazardous chemicals of which students need to be aware. Your instructor will review the salient points on the SDSs for the hazardous chemicals students will be working with and also post the SDSs in the lab for future reference.

Safety Acknowledgment Form: Science Rules and Regulations

I have read the science rules and regulations in the *Student Lab Manual for Argument-Driven Inquiry in Physics, Volume 1,* and I agree to follow them during any science course, investigation, or activity. By signing this form, I acknowledge that the science classroom, laboratory, or field sites can be an unsafe place to work and learn. The safety rules and regulations are developed to help prevent accidents and to ensure my own safety and the safety of my fellow students. I will follow any additional instructions given by my instructor. I understand that I may ask my instructor at any time about the rules and regulations if they are not clear to me. My failure to follow these science laboratory rules and regulations may result in disciplinary action.

_____   _____
Student Signature   Date

_____   _____
Parent/Guardian Signature   Date

National
Science
Teachers
Association

SECTION 2
Space Systems

Introduction Labs

LAB 1

Lab 1. Moon Phases: Why Does the Appearance of the Moon Change Over Time in a Predictable Pattern?

Introduction

We have all seen the Moon in the sky and how it looks different at various times of the month. In fact, differences in the appearance of the Moon over time were the basis for the Chinese, Islamic, Hindu, and Judaic calendars, as well as most of the other calendar systems that were used in ancient times. People can use the appearance of the Moon to mark the passage of time because the Moon's appearance changes in a predictable pattern over a period of 29.5 days. Figure L1.1 shows the pattern that the appearance of the Moon follows. As can be seen in this figure, the portion of the Moon that is illuminated gradually increases until the Moon is full, and then the portion of the Moon that is illuminated gradually decreases until it is completely dark. People often describe this pattern as a lunar cycle. Each phase, or how the Moon looks at a given point in the lunar cycle, has a specific name (see Figure L1.1).

FIGURE L1.1

The phases of the Moon follow a predictable pattern over a period of 29.5 days

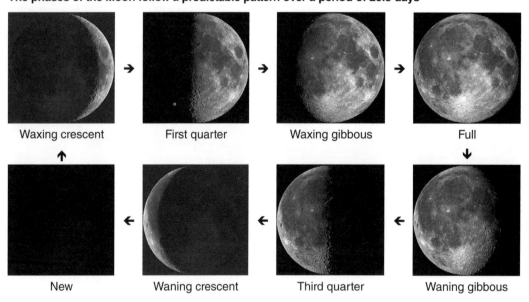

Waxing crescent	First quarter	Waxing gibbous	Full
New	Waning crescent	Third quarter	Waning gibbous

There are some other important facts that we know about Moon in addition to the fact that it goes through a series of phases over the course of a lunar cycle. First, the Moon rises in the east and sets in the west once every 24 hours. The Moon, therefore, travels from east to west across the sky just like the Sun. Second, the time that the Moon rises and sets in the sky changes each day. Sometimes the Moon will rise at dusk and set at dawn, and other times it will rise late at night and set in the morning. The Moon can even rise at dawn and set at dusk just like the Sun. The times that we can see the Moon in the sky therefore change over the course of a lunar cycle. Third, we always see the same light and dark regions on the surface of Moon regardless of its current phase (see Figure L1.1). We always see the same surface features when we look at the Moon because the same side of the Moon is always facing Earth. Finally, we see solar and lunar eclipses from time to time. A solar eclipse occurs during the day. A solar eclipse results in the light from the Sun being blocked for about 5 to 10 minutes (see Figure L1.2). A lunar eclipse, in contrast, occurs at night. A lunar eclipse causes the full Moon to get darker and turn red for a few minutes (see Figure L1.3). All of these different facts about the Moon can be explained if you understand what causes the lunar cycle.

FIGURE L1.2

A solar eclipse

FIGURE L1.3

A lunar eclipse

To explain the lunar cycle and all these different facts about the Moon, it is important to know a little about the types of objects that are found in our solar system and how all these objects move over time in relation to each other. The solar system consists of the Sun, the eight official planets, at least five dwarf planets, more than 130 moons, and numerous small bodies (including comets and asteroids). At the center of the solar system is the Sun. The inner solar system includes the planets Mercury, Venus, Earth, and Mars; the dwarf planet Ceres; and three moons. The outer solar system includes the planets Jupiter, Saturn, Uranus, and Neptune; the four other dwarf planets; and the remaining moons. In our solar system, all the planets and dwarf planets orbit (revolve around) the Sun, and all the moons orbit planets or dwarf planets. All the planets in our solar system travel around the Sun in a counterclockwise direction (when looking down from above the Sun's north pole). All of the planets and dwarf plants, with the exception of Venus, Uranus and Pluto, also spin (or rotate) in a counterclockwise direction.

You can use this information about our solar system to develop a physical model of the Earth-Sun-Moon system. You can then use your physical model to explore how Earth, the Sun, and the Moon move in relation to each other and how the light from the Sun illuminates the Moon as it orbits Earth. You can also use your physical model to determine how different positions of Earth, the Sun, and the Moon in relation to each other affect the appearance of the Moon over time (as seen from Earth). You will then be able to use what you learned about how the Moon and Earth move in relation to each other over time by working with a physical model to create a conceptual model that you can use to explain the lunar cycle.

Your Task

Develop a conceptual model that you can use to explain the phases of the Moon. Your conceptual model must be based on what we know about system and system models, patterns, the objects that are found in our solar system, and how these objects move in relationship to each other. You should be able to use your conceptual model to predict when and where you will be able to see the Moon in the sky during a lunar cycle.

The guiding question of this investigation is, *Why does the appearance of the Moon change over time in a predictable pattern?*

Materials

You may use any of the following materials during your investigation:

Equipment
- Safety glasses or goggles (required)
- Physical model of Earth (large ball on a stand)
- Physical model of the Moon A (small ball on a stand)
- Physical model of the Moon B (small ball on a stick)
- Lamp and lightbulb

Other Resources
- Moon phase calendar A (use to develop your conceptual model)
- Moon phase calendar B (use to test your conceptual model)

Safety Precautions

Follow all normal lab safety rules. In addition, take the following safety precautions:

- Wear sanitized indirectly vented chemical-splash goggles throughout the entire investigation (which includes setup and cleanup).
- Use only GFCI-protected electrical receptacles for lamps to prevent or reduce risk of shock.
- Handle the lamps with care; they can get hot when left on for long periods of time.
- Wash hands with soap and water when done collecting the data and after completing the lab.

Investigation Proposal Required?   ☐ Yes   ☐ No

Getting Started

The first step in developing a conceptual model is to design and carry out an investigation to determine how movement of Earth, the Sun, and the Moon over time results in the Moon looking different from our perspective on Earth. To accomplish this task, you will need to create a physical model of the Earth-Sun-Moon system using the available materials. You can then use this physical model to see how light shines on Earth and the Moon when they are in different positions relative to each other. You can also use this model to test your different ideas about the underlying cause of the Moon phases. As you develop your physical model, be sure to consider the following questions:

- What are the boundaries of the system you are studying?
- What are the components of this system?
- How can you quantitatively describe changes within the system over time?
- What could be causing the pattern that we observe?

Once you have used your physical model to test your ideas about the underlying cause of the Moon phases, your group can use what you learned to develop your conceptual model. A conceptual model is an idea or set of ideas that explains what causes a particular phenomenon in nature. People often use words, images, and arrows to describe a conceptual model. Your conceptual model needs to be able to explain why we see the phases of the Moon in the same pattern. It also needs to be able to explain

- why we see the Moon rise in the east and set in the west,
- why the Moon rises and sets at different times of the day,
- why we see the same side of the Moon regardless of its current phase, and
- why there are occasional solar and lunar eclipses.

The last step in your investigation will be to generate the evidence that you need to convince others that your conceptual model is valid or acceptable. To accomplish this goal, you can use your model to predict when and where the Moon will be in the night sky over the next month. You can also attempt to show how using a different version of your model or making a specific change to a portion of your model would make your model inconsistent with data you have or the facts we know about the Moon. Scientists often make comparisons between different versions of a model in this manner to show that a model is valid or acceptable. If you are able to use your conceptual model to make accurate predictions about the behavior of the Moon over time or you are able show how your conceptual model explains the behavior of the Moon better than other models, then you should be able to convince others that it is valid or acceptable.

LAB 1

Connections to the Nature of Scientific Knowledge and Scientific Inquiry

As you work through your investigation, be sure to think about

- the use of models as tools for reasoning about natural phenomena, and
- how scientists use different methods to answer different types of questions.

Initial Argument

Once your group has finished collecting and analyzing your data, your group will need to develop an initial argument. Your initial argument needs to include a claim, evidence to support your claim, and a justification of the evidence. The claim is your group's answer to the guiding question. The evidence is an analysis and interpretation of your data. Finally, the justification of the evidence is why your group thinks the evidence matters. The justification of the evidence is important because scientists can use different kinds of evidence to support their claims. Your group will create your initial argument on a whiteboard. Your whiteboard should include all the information shown in Figure L1.4.

FIGURE L1.4

Argument presentation on a whiteboard

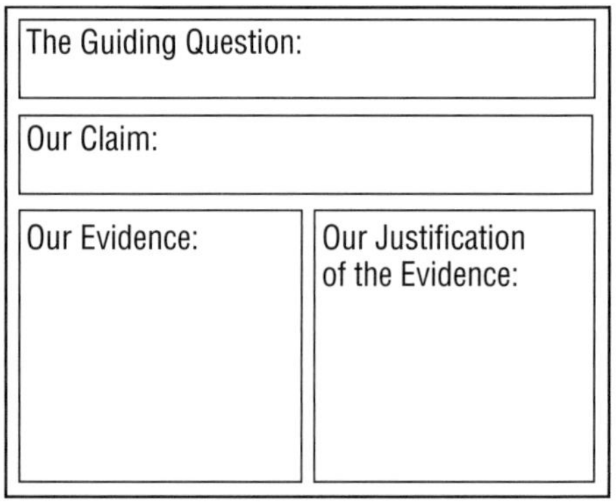

The Guiding Question:	
Our Claim:	
Our Evidence:	Our Justification of the Evidence:

Argumentation Session

The argumentation session allows all of the groups to share their arguments. One or two members of each group will stay at the lab station to share that group's argument, while the other members of the group go to the other lab stations to listen to and critique the other arguments. This is similar to what scientists do when they propose, support, evaluate, and refine new ideas during a poster session at a conference. If you are presenting your group's argument, your goal is to share your ideas and answer questions. You should also keep a record of the critiques and suggestions made by your classmates so you can use this feedback to make your initial argument stronger. You can keep track of specific critiques and suggestions for improvement that your classmates mention in the space below.

Critiques of our initial argument and suggestions for improvement:

If you are critiquing your classmates' arguments, your goal is to look for mistakes in their arguments and offer suggestions for improvement so these mistakes can be fixed. You should look for ways to make your initial argument stronger by looking for things that the other groups did well. You can keep track of interesting ideas that you see and hear during the argumentation session in the space below. You can also use this space to keep track of any questions that you will need to discuss with your team.

Interesting ideas from other groups or questions to take back to my group:

Once the argumentation session is complete, you will have a chance to meet with your group and revise your initial argument. Your group might need to gather more data or design a way to test one or more alternative claims as part of this process. Remember, your goal at this stage of the investigation is to develop the best argument possible.

Report

Once you have completed your research, you will need to prepare an investigation report that consists of three sections. Each section should provide an answer for the following questions:

1. What question were you trying to answer and why?

2. What did you do to answer your question and why?

3. What is your argument?

Your report should answer these questions in two pages or less. You should write your report using a word processing application (such as Word, Pages, or Google Docs), if possible, to make it easier for you to edit and revise it later. You should embed any diagrams, figures, or tables into the document. Be sure to write in a persuasive style; you are trying to convince others that your claim is acceptable or valid.

LAB 1

Checkout Questions

Lab 1. Moon Phases: Why Does the Appearance of the Moon Change Over Time in a Predictable Pattern?

1. The diagram below shows Earth and the Sun, as well as four different possible positions for the Moon (A, B, C, and D).

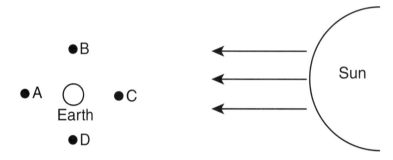

a. Which position of the Moon (A, B, C, or D) would cause it to appear like the picture below when viewed from Earth? Circle the letter below.

A   B   C   D

b. How do you know?

National Science Teachers Association

2. You observe a crescent Moon rising in the east.

a. Which of the following pictures illustrates how it will appear in six hours?

 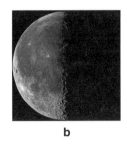  

    a   b   c   d

b. How do you know?

3. One night you look at the Moon and see this:

a. Which of the following pictures illustrates how it will appear one week later?

    a   b   c   d

b. How do you know?

4. The image below is a model of the Earth-Sun-Moon system.

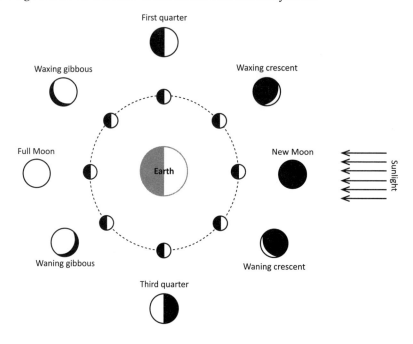

a. What are some strengths of this model?

b. Based on what you know about the Earth-Sun-Moon system from your investigation, what are some limitations or incorrect portions of this model?

5. Scientists often look at patterns to identify relationships in nature. Give an example of a pattern from your investigation on Moon phases.

6. Models are pictures of things that we cannot see.

    a. I agree with this statement.

    b. I disagree with this statement.

    Explain your answer, using an example from your investigation about the phases of the Moon.

7. There is a single scientific method that all scientists follow when conducting an investigation.

   a. I agree with this statement.

   b. I disagree with this statement.

   Explain your answer, using an example from your investigation about phases of the Moon.

Seasons

What Causes the Differences in Average Temperature and the Changes in
Day Length That We Associate With the Change in Seasons on Earth?

Lab Handout

Lab 2. Seasons: What Causes the Differences in Average Temperature and the Changes in Day Length That We Associate With the Change in Seasons on Earth?

Introduction

A season is a subdivision of a year, which is often marked by changes in average daily temperature, amount of precipitation, and hours of daylight. People who live in temperate and subpolar regions around the globe experience four calendar-based seasons: spring, summer, fall, and winter. People who live in regions near the equator, in contrast, only experience two seasons: a rainy (or monsoon) season and a dry season.

Figure L2.1 shows four satellite images of Lake George in New York in February, April, July, and October. These images illustrate how the surface of the Earth looks different during different seasons.

FIGURE L2.1 _____

The change of seasons as seen in four satellite images of Lake George, New York, from the Advanced Spaceborne Thermal Emission and Reflection Radiometer instrument on NASA's Terra spacecraft

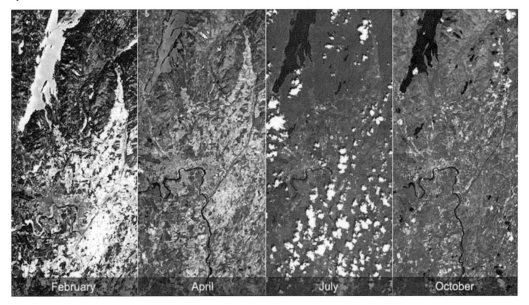

Note: A full-color version of this figure is available on the book's Extras page at *www.nsta.org/adi-ess.*

LAB 2

To understand why we experience different seasons in different locations on Earth, we must first think about the objects that are found in our solar system and how all these objects move over time in relation to each other. The Sun is at the center of our solar system. All the other objects in the solar system, which include planets, dwarf plants, asteroids, and comets, revolve (orbit) around it. All the planets in our solar system travel around the Sun in a counterclockwise direction (when looking down from above the Sun's north pole).

Earth takes 365.25 days to orbit the Sun. The distance that Earth must travel to complete one full revolution around the Sun is 940 million km. Earth, as a result, travels around the Sun at a speed of about 30 km/s. Earth is closest to the Sun in early January due to its slightly elliptical orbit. At this time, the Earth is about 146 million km away from the Sun. Earth is farthest from the Sun in early July, when the distance between Earth and the Sun is about 152 million km.

Earth also spins (or rotates) on its axis as it travels around the Sun. Earth spins on its axis in a counterclockwise direction (when looking down from above Earth's North Pole; see Figure L2.2). It takes 23 hours and 56 minutes for Earth to complete one full rotation. The rotation of Earth on its axis is what gives us day and night. During the day, we are facing the Sun and during the night we are facing away from it. Earth's axis, however, is not perpendicular to its orbit (or straight up if we were able to look down at it from above the solar system). Earth currently has an axial tilt of 23.4° (see Figure L2.2). Earth remains tilted in the same direction regardless of where it is in its orbit. This means that Earth's North Pole is directed toward the Sun in June but directed away from the Sun in December. In contrast, Earth's South Pole is directed toward the Sun in December and is directed away from the Sun in June.

These facts are useful and can help us understand the change in seasons. Yet, these facts do not provide us with all the information that we need to develop a complete conceptual model that explains the cause of the seasons. You will therefore need to learn more about how the average temperature and the hours of daylight change over the course of year at different locations on Earth. Next, you will need to use an online simulation called the Seasons and Ecliptic Simulator to explore how the tilt of Earth affects the amount of sunlight and the angle that sunlight strikes Earth at various locations over time. Finally, you will have an opportunity to put all these pieces of information together to develop a conceptual model that explains the cause of the seasons.

Seasons

What Causes the Differences in Average Temperature and the Changes in Day Length That We Associate With the Change in Seasons on Earth?

FIGURE L2.2

The axial tilt is the angle between a planet's rotational axis at its north pole and a line perpendicular to the orbital plane of the planet. Earth's axial tilt is currently 23.4°.

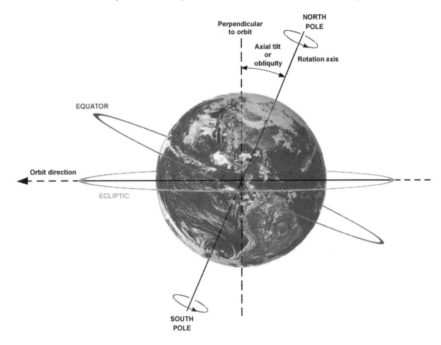

Your Task

Develop a conceptual model that you can use to explain the cause of the seasons. You must base your conceptual model on what we know about how Earth revolves around the Sun and spins on its axis. You will also need to use what you know about systems and system models and the importance of looking for patterns in nature to develop your conceptual model. To be considered valid or acceptable, your conceptual model should not only explain the underlying cause of the seasons but also predict the changes in average daily temperature and hours of daylight at several different locations on Earth.

The guiding question of this investigation is, *What causes the differences in average temperature and the changes in day length that we associate with the change in seasons on Earth?*

Materials

You will use an online simulation called Seasons and Ecliptic Simulator to conduct your investigation; the simulation is available at *http://astro.unl.edu/naap/motion1/animations/ seasons_ecliptic.html*.

Information about the location (latitude and longitude), weather, and hours of daylight for most major cities around the world can be found at *www.climate-charts.com/world-index.html*.

LAB 2

Safety Precautions

Follow all normal lab safety rules.

Investigation Proposal Required?   ☐ Yes   ☐ No

Getting Started

The first step in developing a conceptual model that explains the cause of the seasons is to collect information about the changes in average daily temperature and hours of daylight over a year at several different locations on Earth. This information can be found for cities in 149 countries at the World Climate website, which contains the largest set of accessible climate data on the web. Be sure to collect information from cities at a wide range of latitudes and longitudes. Once you collect this information, look for any patterns that you can use to help develop your conceptual model.

FIGURE L2.3

A screenshot from the Seasons and Ecliptic Simulator simulation

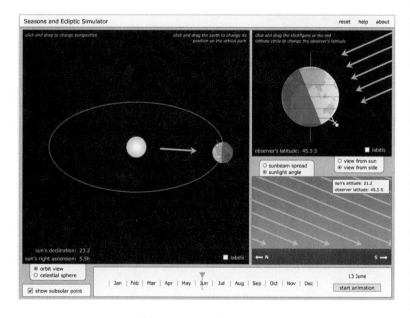

Next, you can use the *Seasons and Ecliptic Simulato*r to learn more about how light from the Sun strikes Earth over the course of the year (see Figure L2.3). This simulation allows you to move an observer to different latitudes on Earth and to track the Sun's altitude in the sky and the sunlight angle over the course of the year for that observer. Given the nature of the simulation, you must determine what type of data you need to collect, how you will collect it, and how you will analyze it to learn more about how Earth's tilt affects the amount of sunlight and the angle at which sunlight strikes Earth over time.

To determine *what type of data you need to collect,* think about the following questions:

- What are the boundaries and components of the system you are studying?
- How do the components of the system interact with each other?
- What type of measurements or observations will you need to record to determine how Earth's tilt affects the amount of sunlight that strikes Earth over time?
- What type of measurements or observations will you need to record to determine how Earth's tilt affects the angle that sunlight strikes Earth over time?

National Science Teachers Association

Seasons

What Causes the Differences in Average Temperature and the Changes in Day Length That We Associate With the Change in Seasons on Earth?

To determine *how you will collect the data,* think about the following questions:

- How often will you need to make the measurements or observations?
- What scale or scales should you use when you take your measurements?
- What types of comparisons will you need to make?

To determine *how you will analyze the data,* think about the following questions:

- What types of calculations will you need to make?
- What types of patterns could you look for as you analyze your data?
- How could you use mathematics to describe a change over time?
- How could you use mathematics to describe a relationship between variables?

Once you have finished using the *Seasons and Ecliptic Simulator,* your group can develop a conceptual model that can be used to explain the cause of the seasons. Be sure to incorporate the information you collected from the World Climate website. To be valid or acceptable, your conceptual model must be able to explain (a) why the length of day changes by different amounts in different locations and (b) why the average temperature for each month changes by different amounts in different locations.

The last step in your investigation will be to generate the evidence that you need to convince others that your conceptual model is valid or acceptable. To accomplish this goal, you can use your model to predict the length of day and average temperature at different times of the year in several additional cities. These cities should be ones that you have not looked up before. You can also attempt to show how using a different version of your model or making a specific change to a portion of your model will make your model inconsistent with data you have or the facts we know about seasons. Scientists often make comparisons between different versions of a model in this manner to show that a model is valid or acceptable. If you are able to use your conceptual model to make accurate predictions about the changes in average daily temperature and hours of daylight at several locations on Earth or you are able show how your conceptual model explains the cause of the seasons better than other models, then you should be able to convince others that it is valid or acceptable.

Connections to the Nature of Scientific Knowledge and Scientific Inquiry

As you work through your investigation, be sure to think about

- the use of models as tools for reasoning about natural phenomena, and
- how scientists use different methods to answer different types of questions.

LAB 2

Initial Argument

Once your group has finished collecting and analyzing your data, your group will need to develop an initial argument. Your initial argument needs to include a claim, evidence to support your claim, and a justification of the evidence. The *claim* is your group's answer to the guiding question. The *evidence* is an analysis and interpretation of your data. Finally, the *justification* of the evidence is why your group thinks the evidence matters. The justification of the evidence is important because scientists can use different kinds of evidence to support their claims. Your group will create your initial argument on a whiteboard. Your whiteboard should include all the information shown in Figure L2.4.

FIGURE L2.4
Argument presentation on a whiteboard

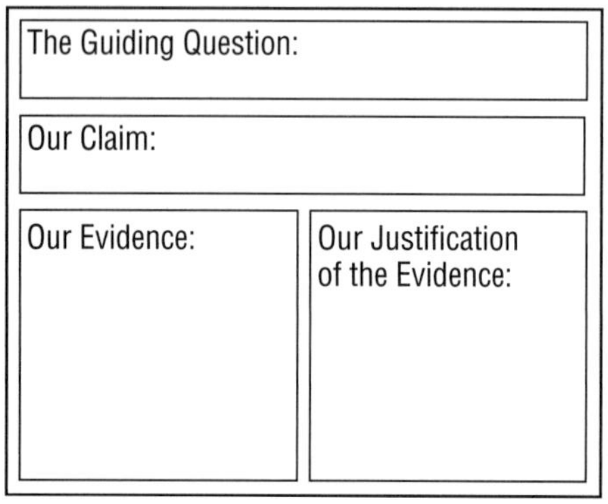

The Guiding Question:	
Our Claim:	
Our Evidence:	Our Justification of the Evidence:

Argumentation Session

The argumentation session allows all of the groups to share their arguments. One or two members of each group will stay at the lab station to share that group's argument, while the other members of the group go to the other lab stations to listen to and critique the other arguments. This is similar to what scientists do when they propose, support, evaluate, and refine new ideas during a poster session at a conference. If you are presenting your group's argument, your goal is to share your ideas and answer questions. You should also keep a record of the critiques and suggestions made by your classmates so you can use this feedback to make your initial argument stronger. You can keep track of specific critiques and suggestions for improvement that your classmates mention in the space below.

Critiques of our initial argument and suggestions for improvement:

Seasons

What Causes the Differences in Average Temperature and the Changes in Day Length That We Associate With the Change in Seasons on Earth?

If you are critiquing your classmates' arguments, your goal is to look for mistakes in their arguments and offer suggestions for improvement so these mistakes can be fixed. You should look for ways to make your initial argument stronger by looking for things that the other groups did well. You can keep track of interesting ideas that you see and hear during the argumentation in the space below. You can also use this space to keep track of any questions that you will need to discuss with your team.

Interesting ideas from other groups or questions to take back to my group:

Once the argumentation session is complete, you will have a chance to meet with your group and revise your initial argument. Your group might need to gather more data or design a way to test one or more alternative claims as part of this process. Remember, your goal at this stage of the investigation is to develop the best argument possible.

Report

Once you have completed your research, you will need to prepare an *investigation report* that consists of three sections. Each section should provide an answer for the following questions:

1. What question were you trying to answer and why?

2. What did you do to answer your question and why?

3. What is your argument?

Your report should answer these questions in two pages or less. You should write your report using a word processing application (such as Word, Pages, or Google Docs), if possible, to make it easier for you to edit and revise it later. You should embed any diagrams, figures, or tables into the document. Be sure to write in a persuasive style; you are trying to convince others that your claim is acceptable or valid.

Checkout Questions

Lab 2. Seasons: What Causes the Differences in Average Temperature and the Changes in Day Length That We Associate With the Change in Seasons on Earth?

1. Pictured below are five different pictures (A–E) of Earth as it orbits the Sun. Use numbers to rank the pictures from highest to lowest temperature at the locations indicated by the black circle. If you think two or more pictures show locations with equal temperatures, give them the same number.

Picture	Relative position of Earth as it orbits the Sun*	Rank
A		_____
B		_____
C		_____
D		_____
E		_____

* Not to scale

Seasons

*What Causes the Differences in Average Temperature and the Changes in
Day Length That We Associate With the Change in Seasons on Earth?*

Explain your answer. Why do you think the order that you chose is correct?

2. The December solstice occurs when Earth is positioned as it is in the picture
   below. Sometimes people refer to this as the winter solstice—is this an appropriate
   name for this solstice?

   Explain your answer, using an example from your investigation about the cause of
   the seasons.

3. Scientists create pictures of things to teach people about them. These pictures are models.

    a. I agree with this statement.
    b. I disagree with this statement.

    Explain your answer, using an example from your investigation about the cause of the seasons.

4. All scientific investigations are experiments.

    a. I agree with this statement.
    b. I disagree with this statement.

    Explain your answer, using an example from your investigation about the cause of the seasons.

Seasons

What Causes the Differences in Average Temperature and the Changes in Day Length That We Associate With the Change in Seasons on Earth?

5. Scientists often work to understand the cause of patterns we observe in nature. Provide an example of a pattern and its cause from your investigation about the cause of the seasons.

6. Scientists often develop or use models. Explain why scientists might use a model to study natural phenomena, using an example from your investigation about the cause of the seasons.

Lab Handout

Lab 3. Gravity and Orbits: How Does Changing the Mass and Velocity of a Satellite and the Mass of the Object That It Revolves Around Affect the Nature of the Satellite's Orbit?

Introduction

The motion of an object is the result of all the different *forces* that act on it. If you pull on a door, the door will move in the direction that you pulled it. If you push on a marble that is resting on a table, the marble will move in the direction you pushed it. Pulling on a door and pushing on a marble are examples of a *contact force*, which is a force that is applied to an object through direct contact. There are other types of forces that can push or pull on an object without touching it. A magnet, for example, can pull or push on another magnet without touching it. Static electricity, which is the buildup of electrical charge on an object, can also pull or push on an object. Magnetic and electrical forces are therefore called *non-contact forces* because they act at a distance. Perhaps the most common non-contact force is *gravity*. Gravity is a force of attraction between two objects; the force due to gravity always works to bring objects closer together.

Any two objects, as long as they have some mass, will have a gravitational force of attraction between them. The strength or magnitude of the gravitational force that exists between any two objects is influenced by the masses of those two objects and the distance between them. The magnitude of gravitational attraction increases with greater mass. This means that the gravitational force that exists between Earth and a car is greater than the gravitational force that exists between Earth and a marble. The magnitude of gravitational attraction, however, decreases as the distance between any two objects increases. The magnitude of the gravitational force that exists between Earth and an object that is moving away from it will therefore get weaker and weaker as the objects moves farther and farther away from Earth.

The force of gravity keeps planets orbiting a star and moons orbiting planets. An *orbit* is a regular, repeating path that one object in space takes around another one. An object in an orbit is called a *satellite*. A satellite can be natural, like planets, moons, and comets, or it can be something that was created by engineers and scientists, such as the International Space Station or the Hubble Space Telescope.

All orbits are *elliptical,* which means that the satellite follows a path that is round but can range in shape from a perfect circle to a long, thin oval. The shape of the orbit that most of the inner planets of our solar system follow, for example, is nearly circular. Figure L3.1 shows the orbits of Venus, Earth, and Mars. Notice that these orbits look almost like perfect circles. The orbits of comets and some of the outer dwarf plants have a very different shape. They are highly *eccentric*. In other words, their orbits look like a squashed circle.

Gravity and Orbits

How Does Changing the Mass and Velocity of a Satellite and the Mass of the Object That It Revolves Around Affect the Nature of the Satellite's Orbit?

FIGURE L3.1

The orbits of Venus, Earth, and Mars as they would appear to an observer located above our solar system (the diagram is not to scale)

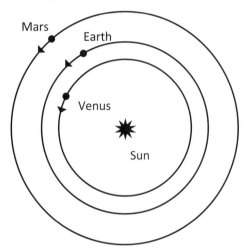

One way to describe the shape of an orbit is to calculate its *eccentricity*. Eccentricity is a way to quantify how much an orbit differs from a perfect circle. It is a value that ranges from 0 to 1. An orbit with an eccentricity of 0 is a perfect circle. Figure L3.2 illustrates orbits with eccentricity values of 0., 0.5, 0.75, and 0.9. The formula used to calculate eccentricity is

$$e = (\sqrt{a^2 - b^2})/a$$

where *e* is the eccentricity of the ellipse, *a* = is the major axis of the ellipse, and *b* = is the minor axis of the ellipse. Figure L3.3 shows how to calculate the eccentricity of an orbit. In this example, the major axis of the ellipse (*a*) is 7 units long and the minor axis (*b*) is 5 units long. Substituting these values into the formula gives a value of 0.7. This elliptical orbit would be considered highly eccentric.

FIGURE L3.2

Orbits with different eccentricities

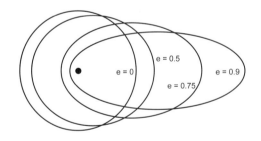

FIGURE L3.3

How to calculate the eccentricity of an orbit

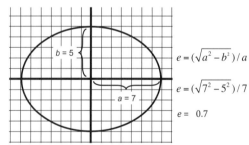

Another way to describe the orbit of a satellite is to measure its orbital distance. Satellites, however, do not always stay the same distance from the star or planet that they orbiting because their orbits are elliptical. For planets, like Earth, the point in their orbit when they are closest to the Sun is called the perihelion (see Figure L3.4). The point where a planet is farthest from the Sun is called the aphelion. The closest point the Moon or a manufactured satellite comes to Earth is called its perigee, and the farthest point is the apogee. Earth reaches its aphelion during July and its perihelion in January. The third, and final, way to describe the orbit of a satellite is to measure the time it takes to make one full orbit. The amount of time required to complete an orbit is called the orbital period. Earth, for example, has an orbital period of one year.

LAB 3

In this investigation, you will have an opportunity to use an online simulation to explore how three different factors affect the shape, distance, and period of a satellite's orbit. The first factor is the mass of the satellite. The second factor is its initial velocity (speed in a given direction). The third factor is the mass of the object that it is orbiting. This type of investigation can be difficult because identifying the exact nature of the relationship that exists between multiple factors is challenging. Take mass as an example. There are many potential ways that the mass of a satellite or the mass of the object it is orbiting could influence the satellite's orbit. The shape, distance, and period of the orbit may depend on the mass of the larger object and/or the mass of the smaller object. The mass of the satellite and the mass of the object it is orbiting could also change these three aspects of a satellite's orbit in different ways.

FIGURE L3.4

Illustration of the orbit of Earth around the Sun and of the Moon around Earth showing the aphelion, perihelion, perigee, and apogee. (The orbits of the Earth and the Moon are not as eccentric as they appear in this image.)

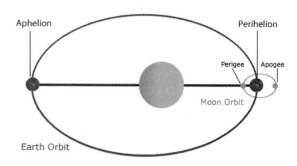

In addition to mass, there are many different ways that an orbit might change due to a change in the initial velocity of a satellite. The eccentricity of an orbit may either increase or decrease as the initial velocity increases. The initial velocity may also affect the orbital period but may not change the distance of its perigee and apogee (or perihelion and aphelion if the satellite is a planet). All of these different relationships are possible, as well as many others. Your goal in this investigation is to determine how all three factors are related to each other so you can better understand and predict the shape, distance, and period of a satellite's orbit.

Your Task

Use what you know about gravity; scale, proportion, and quantity; and the role of models in science to design and carry out an investigation that will allow you to determine how three different factors affect the shape, distance, and period of a satellite's orbit. The three factors you will explore are the mass of the satellite, the initial velocity of the satellite, and the mass of the object that the satellite is orbiting.

The guiding question of this investigation is, *How does changing the mass and velocity of a satellite and the mass of the object that it revolves around affect the nature of the satellite's orbit?*

Materials

You will use an online simulation called *My Solar System* to conduct your investigation; the simulation is available at *https://phet.colorado.edu/en/simulation/legacy/my-solar-system.*

National Science Teachers Association

Gravity and Orbits

How Does Changing the Mass and Velocity of a Satellite and the Mass of the Object
That It Revolves Around Affect the Nature of the Satellite's Orbit?

Safety Precautions

Follow all normal lab safety rules.

Investigation Proposal Required?   ☐ Yes   ☐ No

Getting Started

The *My Solar System* simulation (see Figure L3.5) enables you to observe the orbit of a planet as it orbits around a star. It also allows you to change the mass of the planet and the star to see how changes in mass affects the shape, distance, and period of the planet's orbit. You can also add additional bodies to the solar system and change the initial velocity of any object that is orbiting the star.

FIGURE L3.5 _____

A screenshot from the *My Solar System* simulation

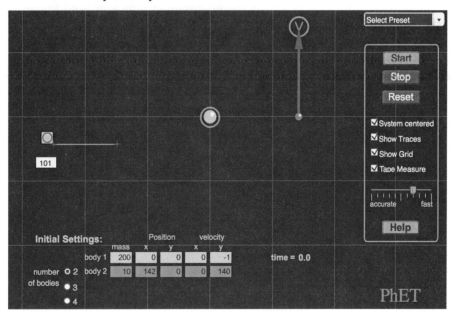

To use this simulation, start by making sure that the boxes next to System Centered, Show Traces, Show Grid, and Tape Measure in the control panel on the right side of the screen are all checked. This will make it easier for you to take the measurements. You can add or remove bodies from the solar system by clicking on the radio buttons in the lower left corner. The mass, initial position, and initial velocity of each body in the solar system can also be changed by typing in new values for each factor using the text boxes at the bottom of the simulation. This simulation is useful because it allows you see the path a planet takes as it orbits a star, and perhaps more important, it provides a way for you to design and carry out controlled experiments. This is important because you must be able to manipulate variables during a controlled experiment, and many of the variables that we are interested in here, such as the mass of a star, the mass of a planet, or the initial velocity of a planet, cannot be changed in the real world.

You will need to design and carry out at least three different experiments using the *My Solar System* simulation to determine the relationship between the three factors and the nature of a satellite's orbit. Remember, any object in an orbit is called a satellite. A satellite can be natural, like planets and moons, or a satellite can something that is manufactured and sent into space. You will need to conduct three different experiments because you will need to be able to answer three specific questions before you will be able to develop an answer to the guiding question for this lab:

- How does changing the mass of the star affect the way a planet orbits around it?
- How does changing the mass of a planet affect the way it orbits around a star?
- How does changing the velocity of a planet affect the way it orbits around a star?

It will be important for you to determine what type of data you need to collect, how you will collect the data, and how you will analyze the data for each experiment because each experiment is slightly different. To determine *what type of data you need to collect*, think about the following questions:

- What are the components of this system and how do they interact?
- How can you describe the components of the system quantitatively?
- What information will you need to determine the perihelion and aphelion (or perigee and apogee) of an orbit during each experiment?
- What information will you need to calculate the eccentricity of an orbit during each experiment?
- What information will you need to determine an orbital period during each experiment?

To determine *how you will collect the data*, think about the following questions:

- What will serve as your independent and dependent variable for each experiment?
- How will you vary the independent variable during each experiment?
- What will you do to hold the other variables constant during each experiment?
- When will you need to take measurements or observations during each experiment?
- What scale or scales should you use when you take your measurements?
- What types of comparisons will you need to make using the simulation?
- How will you keep track of the data you collect and how will you organize it?

Gravity and Orbits

*How Does Changing the Mass and Velocity of a Satellite and the Mass of the Object
That It Revolves Around Affect the Nature of the Satellite's Orbit?*

To determine *how you will analyze the data,* think about the following questions:

- How will you compare the perihelion and aphelion (or perigee and apogee) of an orbit?
- How will you calculate the eccentricity of an orbit?
- How will you compare the eccentricities of several different orbits?
- How will you determine an orbital period?
- How will you compare the periods of several different orbits?
- What potential proportional relationships can you find in the data?

Once you have carried out all your different experiments, your group will need to develop an answer to the guiding question for this investigation. To be sufficient, your answer must explain how the mass of the satellite, the initial velocity of the satellite, and the mass of the object that the satellite is orbiting affect the eccentricity, the perihelion and aphelion, and the period of an orbit. For it to be valid and acceptable, your answer will also need to be consistent with your findings from all three experiments.

Connections to the Nature of Scientific Knowledge and Scientific Inquiry

As you work through your investigation, be sure to think about

- the difference between laws and theories in science, and
- the assumptions made by scientists about order and consistency in nature.

Initial Argument

Once your group has finished collecting and analyzing your data, your group will need to develop an initial argument. Your initial argument needs to include a claim, evidence to support your claim, and a justification of the evidence. The claim is your group's answer to the guiding question. The evidence is an analysis and interpretation of your data. Finally, the justification of the evidence is why your group thinks the evidence matters. The justification of the evidence is important because scientists can use different kinds of evidence to support their claims. Your group will create your initial argument on a whiteboard. Your whiteboard should include all the information shown in Figure L3.6.

FIGURE L3.6 _____

Argument presentation on a whiteboard

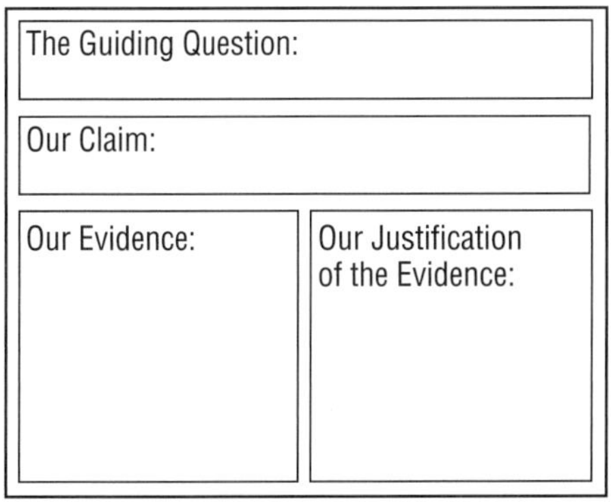

The Guiding Question:	
Our Claim:	
Our Evidence:	Our Justification of the Evidence:

Argumentation Session

The argumentation session allows all of the groups to share their arguments. One or two members of each group will stay at the lab station to share that group's argument, while the other members of the group go to the other lab stations to listen to and critique the other arguments. This is similar to what scientists do when they propose, support, evaluate, and refine new ideas during a poster session at a conference. If you are presenting your group's argument, your goal is to share your ideas and answer questions. You should also keep a record of the critiques and suggestions made by your classmates so you can use this feedback to make your initial argument stronger. You can keep track of specific critiques and suggestions for improvement that your classmates mention in the space below.

Critiques of our initial argument and suggestions for improvement:

If you are critiquing your classmates' arguments, your goal is to look for mistakes in their arguments and offer suggestions for improvement so these mistakes can be fixed. You should look for ways to make your initial argument stronger by looking for things that the other groups did well. You can keep track of interesting ideas that you see and hear during the argumentation in the space below. You can also use this space to keep track of any questions that you will need to discuss with your team.

Interesting ideas from other groups or questions to take back to my group:

Gravity and Orbits

How Does Changing the Mass and Velocity of a Satellite and the Mass of the Object That It Revolves Around Affect the Nature of the Satellite's Orbit?

Once the argumentation session is complete, you will have a chance to meet with your group and revise your initial argument. Your group might need to gather more data or design a way to test one or more alternative claims as part of this process. Remember, your goal at this stage of the investigation is to develop the best argument possible.

Report

Once you have completed your research, you will need to prepare an investigation report that consists of three sections. Each section should provide an answer for the following questions:

1. What question were you trying to answer and why?

2. What did you do to answer your question and why?

3. What is your argument?

Your report should answer these questions in two pages or less. You should write your report using a word processing application (such as Word, Pages, or Google Docs), if possible, to make it easier for you to edit and revise it later. You should embed any diagrams, figures, or tables into the document. Be sure to write in a persuasive style; you are trying to convince others that your claim is acceptable or valid.

Checkout Questions

Lab 3. Gravity and Orbits: How Does Changing the Mass and Velocity of a Satellite and the Mass of the Object That It Revolves Around Affect the Nature of the Satellite's Orbit?

1. How is an object's mass related to the force of gravity it will exert on other objects?

2. How does the gravitational force of an object change as distance from the object increases?

3. A scientist has drawn a sketch of four planet-satellite pairs. The sizes of the circles represent the relative masses of the objects, and the size and direction of the arrows represent the initial velocity and direction of the satellites.

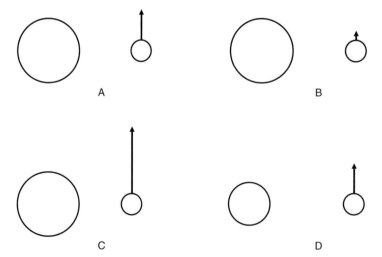

Gravity and Orbits

How Does Changing the Mass and Velocity of a Satellite and the Mass of the Object
That It Revolves Around Affect the Nature of the Satellite's Orbit?

a. Which pair will most likely result in the satellite not entering a stable orbit?

A   B   C   D

b. How do you know?

c. Which pair will most likely result in the satellite crashing into the planet?

A   B   C   D

d. How do you know?

4. Theories are guesses and laws are facts.

    a. I agree with this statement.

    b. I disagree with this statement.

Explain your answer, using an example from your investigation about gravity and orbits.

5. Scientists assume that the universe is a vast single system in which basic laws are consistent.

   a. I agree with this statement.
   b. I disagree with this statement.

   Explain your answer, using an example from your investigation about gravity and orbits.

6. Scientists often use models to represent the components of a system and how these components interact with each other. Explain why models of systems are useful, using an example from your investigation about gravity and orbits.

7. It is critical for scientists to be able to keep track of changes in a system quantitatively during an investigation. Explain why this is so important, using an example from your investigation about gravity and orbits.

Application Lab

LAB 4

Lab 4. Habitable Worlds: Where Should NASA Send a Probe to Look for Life?

Introduction

Our solar system consists of the star we call the Sun, the planets and dwarf plants that orbit it, and the moons that orbit the planets or dwarf planets. It also contains much smaller objects such as comets, asteroids, and dust. The Sun is one of many stars that have several objects orbiting around it in our Milky Way galaxy. As far as we know, Earth is the only planet that supports life. Earth has several unique properties that allow life to exist; it is solid, it is warm enough to allow liquid water, and it is not too cold to freeze its inhabitants. However, we have only extensively studied the planets and dwarf planets within our own solar system. We know some information about planets outside of our solar system, called *exoplanets,* but exoplanets are difficult to study because they are so far away.

Space scientists with the National Aeronautics and Space Administration (NASA) are currently looking for exoplanets as part of a long-term project called Kepler. To find exoplanets, space scientists aim powerful telescopes at stars outside of our solar system and then measure the brightness of the light that the star emits over long periods of time. They then look for the brightness of the star to dim in a pattern because it indicates that an object is orbiting that star. Figure L4.1 shows what happens to the brightness of a star over time as a planet moves around it. The top row shows what a star would look like to us from Earth as a planet orbits around it. The second row shows how the brightness of the star remains constant except when the planet crosses in front of it—at those moments, the brightness of the star decreases. Kepler space scientists describe

FIGURE L4.1

An example of how Kepler space scientists identify exoplanets. The top panels show how the position of a hypothetical exoplanet changes over time as it orbits a hypothetical star when viewed from Earth. The bottom graph shows how the brightness of the star will decrease only when the exoplanet passes between the star and the Earth.

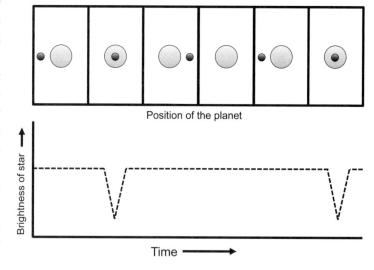

this instance as a transit event, and they use computers to identify potential transit events. These potential transit events are classified as *Kepler objects of interest* (KOIs). Kepler space scientists then investigate each KOI to determine if the change in brightness of a star was caused by an exoplanet or something else. So far, space scientists working on the Kepler project have identified over 4,000 exoplanets using this method.

Once the Kepler space scientists have identified an exoplanet, they attempt to determine if it can support life. To accomplish this task, the space scientists look to see if the exoplanet shares three important characteristics with Earth:

1. It must be terrestrial (composed primarily of rocks and metals).

2. It must have an atmosphere.

3. It must orbit within the habitable zone around a star, which is the minimum and maximum distance from a star where the temperature of a planetary surface can support liquid water given sufficient atmospheric pressure.

Space scientists can determine if an exoplanet is terrestrial or not based on its size, because any exoplanet that is more than twice the size of Earth is likely to be a gas giant. Space scientists can also use the size of an exoplanet to determine if it has an atmosphere or not. Any exoplanet that is less than half the size of Earth most likely does not have enough gravitational pull to keep an atmosphere around it. Finally, space scientists can determine if an exoplanet orbits a star within the habitable zone by measuring its orbital period. The location of the habitable zone around a star, however, is not the same for every exoplanet because some stars emit more energy than others.

These three characteristics are important because they are needed to be able to sustain liquid water on the surface of the exoplanet, and liquid water must be present to support life as we know it. Currently, the Kepler space scientists have identified 12 exoplanets that have all three of these characteristics, so it is possible that these 12 planets could support life.

Scientists are interested in finding exoplanets with life on them because learning more about the life found on these exoplanets will help us better understand the evolution of life on Earth. Information about the characteristics of life on exoplanets could either corroborate existing theories about how life began and changes over time or result in new theories. However, it would cost too much money and waste too much time to send exploratory probes to every exoplanet they find to look for life. So scientists must be able to identify the exoplanets with the highest chance of supporting life. In this investigation, you will have an opportunity to use data about several different exoplanets to determine which ones, if any, have the potential to support life.

Your Task

Use what you know about the Earth-Sun system, patterns, and how scientists need to consider different scales, proportional relationships, and quantities during an investigation to examine the characteristics of several exoplanets orbiting around distant stars. Your goal is to determine which exoplanet, if any, is most likely to contain life based on its physical properties, the properties of the star it orbits, and the size and shape of its orbit.

LAB 4

The guiding question of this investigation is, ***Where should NASA send a probe to look for life?***

Materials

You will use the Lab 4 Reference Sheet: Kepler Project Information Packet during your investigation.

Safety Precautions

Follow all normal lab safety rules.

Investigation Proposal Required?   ☐ Yes   ☐ No

Getting Started

To answer the guiding question, you will need to analyze an existing data set. To determine *how you will analyze the data,* think about the following questions:

- Which data are relevant based on the guiding question?
- What type of calculations will you need to make?
- What types of patterns might you look for as you analyze your data?
- Are there any proportional relationships that you can identify?
- How will you determine if the physical properties of the stars and their planet candidates are the same or different?
- How could you use mathematics to determine if there is or is not a difference?
- What type of table or graph could you create to help make sense of your data?

Connections to the Nature of Scientific Knowledge and Scientific Inquiry

As you work through your investigation, be sure to think about

- the difference between data and evidence in science, and
- the assumptions made by scientists about order and consistency in nature.

Initial Argument

Once your group has finished collecting and analyzing your data, your group will need to develop an initial argument. Your initial argument needs to include a claim, evidence to support your claim, and a justification of the evidence. The *claim* is your group's answer to the guiding question. The *evidence* is an analysis and interpretation of your data. Finally, the *justification* of the evidence is why your group thinks the evidence matters. The justification of the evidence is important because scientists can use different kinds of evidence

FIGURE L4.2 _____

Argument presentation on a whiteboard

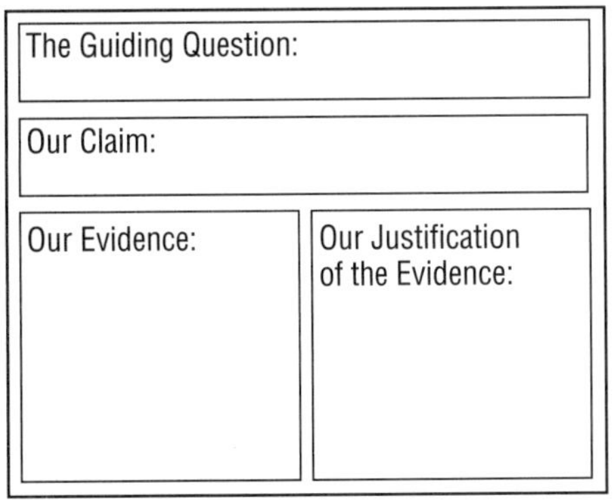

The Guiding Question:	
Our Claim:	
Our Evidence:	Our Justification of the Evidence:

to support their claims. Your group will create your initial argument on a whiteboard. Your whiteboard should include all the information shown in Figure L4.2.

Argumentation Session

The argumentation session allows all of the groups to share their arguments. One or two members of each group will stay at the lab station to share that group's argument, while the other members of the group go to the other lab stations to listen to and critique the other arguments. This is similar to what scientists do when they propose, support, evaluate, and refine new ideas during a poster session at a conference. If you are presenting your group's argument, your goal is to share your ideas and answer questions. You should also keep a record of the critiques and suggestions made by your classmates so you can use this feedback to make your initial argument stronger. You can keep track of specific critiques and suggestions for improvement that your classmates mention in the space below.

Critiques about our initial argument and suggestions for improvement:

If you are critiquing your classmates' arguments, your goal is to look for mistakes in their arguments and offer suggestions for improvement so these mistakes can be fixed. You should look for ways to make your initial argument stronger by looking for things that the other groups did well. You can keep track of interesting ideas that you see and hear during the argumentation in the space below. You can also use this space to keep track of any questions that you will need to discuss with your team.

Interesting ideas from other groups or questions to take back to my group:

Once the argumentation session is complete, you will have a chance to meet with your group and revise your initial argument. Your group might need to gather more data or design a way to test one or more alternative claims as part of this process. Remember, your goal at this stage of the investigation is to develop the best argument possible.

Report

Once you have completed your research, you will need to prepare an *investigation report* that consists of three sections. Each section should provide an answer for the following questions:

1. What question were you trying to answer and why?

2. What did you do to answer your question and why?

3. What is your argument?

Your report should answer these questions in two pages or less. You should write your report using a word processing application (such as Word, Pages, or Google Docs), if possible, to make it easier for you to edit and revise it later. You should embed any diagrams, figures, or tables into the document. Be sure to write in a persuasive style; you are trying to convince others that your claim is acceptable or valid.

Checkout Questions

Lab 4. Habitable Worlds: Where Should NASA Send a Probe to Look for Life?

1. Draw a model showing the relationship between Earth, the Sun, an exoplanet, its star, the Milky Way galaxy, and the universe.

2. What are three characteristics that make life possible on Earth?

3. An exoplanet is discovered that has the following characteristics:

   - A diameter 1.7 times that of Earth
   - An orbital period of 42.6 days
   - An orbit semi-major axis of 0.8823 AU

   a. Based on the information provided, how confident can you be that this exo-planet is able to support life as we know it? Mark your answer on the line below

   Not at all confident • _____ • Very confident

   b. Explain your answer.

   c. What additional information would make you more confident and why?

4. A list of planet-star radius ratios is an example of evidence.

   a. I agree with this statement.

   b. I disagree with this statement.

   Explain your answer, using an example from your investigation about habitable worlds.

5. Scientists assume the universe is a vast single system in which basic laws are consistent.

   a. I agree with this statement.

   b. I disagree with this statement.

   Explain your answer, using an example from your investigation about habitable worlds.

6. Scientists often need to look for patterns that occur in the data they collect and analyze. Explain why identifying patterns is important, using an example from your investigation about habitable worlds.

7. It is critical for scientists to be able to describe components of a system quantitatively. Explain why it is important to be able to describe a system quantitatively, using an example from your investigation about habitable worlds.

SECTION 3
History of Earth

Introduction Labs

Lab Handout

Lab 5. Geologic Time and the Fossil Record: Which Time Intervals in the Past 650 Million Years of Earth's History Are Associated With the Most Extinctions and Which Are Associated With the Most Diversification of Life?

Introduction

Earth scientists use the structure, sequence, and properties of rocks, sediments, and fossils, as well as the locations of current and past ocean basins, lakes, and rivers, to learn about the major events in Earth's history. Major historical events include the formation of mountain chains and ocean basins, volcanic eruptions, periods of massive glaciation (when ice glaciers increase in size because of colder than average global temperatures), the development of watersheds and rivers, and the evolution and extinction of different types of organisms. Earth scientists can determine when and where these major events happened because rock layers, such as the ones pictured in Figure L5.1, provide a lot of information about how an area has changed over time. Earth scientists can also determine when these changes happened by determining the *absolute age* or *relative age* of different layers. Earth scientists can determine the absolute age of a layer of rock by measuring the amount of different radioactive elements found in a layer, and they can determine the relative ages of different rock layers using some fundamental ideas about the ways layers of rock form over time that are based on our understanding of geologic processes.

FIGURE L5.1 _____

Horizontal rock layers are easy to see (a) at the Grand Canyon in Arizona and (b) near Khasab in Oman (a country in the Middle East)

a   b

Geologic Time and the Fossil Record

Which Time Intervals in the Past 650 Million Years of Earth's History Are Associated With the Most Extinctions and Which Are Associated With the Most Diversification of Life?

There are four fundamental ideas that Earth scientists use to study and determine the relative age of rock layers. Nicholas Steno introduced the first two in 1669 (see *www.ucmp. berkeley.edu/history/steno.html*). The first fundamental idea, which is called the *principle of original horizontality,* states that sedimentary rocks are originally laid down in horizontal layers; see Figure L5.1 for an example of this principle. The second fundamental idea is called the *law of superposition.* This law states that in an undisturbed column of rock, the youngest rocks are at the top and the oldest are at the bottom. The third fundamental idea is known as the *principle of uniformitarianism;* it was introduced by James Hutton in 1785, and later expanded by Charles Lyell in the early 1800s (see *www.uniformitarianism. net*). This idea states that geologic processes are consistent throughout time. William Smith introduced the fourth fundamental idea in 1816. This idea is called the *principle of faunal succession,* which states that fossils are found in rocks in a specific order (see *https://earthob-servatory.nasa.gov/Features/WilliamSmith/page2.php*). This principle led Earth scientists to use fossils as a way to define increments of time within the geologic time scale. Because many individual plant and animal species existed during known time periods, the location of certain types of fossils in a rock layer can reveal the age of the rocks and help Earth scientists decipher the history of landforms.

The geologic history of the Earth, or geologic time scale, is broken up into hierarchical chunks of time based on the major events in Earth's history. These major events can be catastrophic, occurring over hours to years, or gradual, occurring over thousands to millions of years. Records of fossils and other rocks also show past periods of massive extinctions and extensive volcanic activity. From largest to smallest, this hierarchical organization of the geological time based on the major events includes eons, eras, periods, epochs, and stages. All of these are displayed in the portion of the geologic time scale shown in Table L5.1.

The majority of macroscopic organisms (organisms that can be seen by the human eye without a microscope), have lived during the Phanerozoic eon; these organisms include algae, fungi, plants, and animals. When Earth scientists first proposed the Phanerozoic eon as a division of geologic time, they believed that the beginning of this eon (542 million years ago [mya]) marked the beginning of life on Earth. We now know that this eon only marks the appearance of macroscopic organisms and that life on Earth actually began about 3.8 billion years ago as single-celled organisms. The Phanerozoic is subdivided into three major divisions, called eras: the Cenozoic, the Mesozoic, and the Paleozoic (see the "Geologic Time Scale" web page at *www.ucmp.berkeley.edu/help/timeform.php*). The suffix *zoic* means animal. The root *Ceno* means recent, the root *Meso* means middle, and the root *Paleo* means ancient. These divisions mark major changes in the nature or composition of life found on Earth. For example, the Cenozoic (65.5 mya–present) is sometimes called the age of mammals because the largest animals on Earth during this era have been mammals, whereas the Mesozoic (251–65.5 mya) is sometimes called the age of dinosaurs because these animals were found on Earth during this time period. The Paleozoic (542–251 mya), in contrast, is divided into six different periods that mark the appearance of different kinds of invertebrate and vertebrate animals. The descriptions of these eras and periods can be somewhat misleading,

LAB 5

TABLE L5.1

Some of the eons, eras, periods, epochs, and stages in the geologic time scale, based on data from Gradstein, Ogg, and Hilgen (2012). Notice that different stages are nested within an epoch, different epochs are nested within a period, and different periods are nested within an era.

Eon	Era	Period	Epoch	Stage	Time (mya)
Phanerozoic	Cenozoic	Neogene	Pliocene	Piacenzian	2.6-3.6
				Zanclean	3.6-5.3
			Miocene	Messinian	5.3-7.3
				Tortonian	7.3-11.6
				Serravalian	11.6-13.8
				Langhian	13.8-15.9
				Burdigalian	15.9-20.4
				Aquitanian	20.4-23.0
		Paleogene	Oligocene	Chattian	23.0-28.1
				Rupelian	28.1-33.9
			Eocene	Priabonian	33.9-37.8
				Bartonian	37.8-41.2
				Lutetian	41.2-47.8
				Ypresian	47.8-56.0

however, because many different groups of animals lived during each of them. There were also many kinds of plants living during these different eras and periods.

Earth scientists have collected a lot of data about the history of life on Earth over the last 400 years. This information not only allows scientists to determine what the conditions were like on Earth in the past, but also allows them to track when major groups of animals and plants appeared and disappeared during the past 650 million years of Earth's history. In this investigation, you will have an opportunity to learn about the extinction and diversification of life on Earth. It is important to note, however, that the fossil record only provides a partial picture how life on Earth has changed over time. Although it is substantial, the fossil record is incomplete because life forms that are common, widespread, and have hard shells or skeletons are more likely to be preserved as fossils than life forms that are rare, isolated, and have soft bodies. The fossil record, therefore, can only provide limited information about the history of life on Earth.

Your Task

Use a database called The Fossil Record 2 and what you know about geologic time, patterns, scales, proportions, and quantities to identify any major extinction and diversification events that happened during the last 650 million years. You goal is to determine how many of each of these important events occurred and when they happened in the geologic time scale.

The guiding question of this investigation is, *Which time intervals in the past 650 million years of Earth's history are associated with the most extinctions and which are associated with the most diversification of life?*

Materials

You will use a computer with Excel or other spreadsheet application during your investigation. You will also use the following resources:

- Fossil Record 2 Database Summary Counts Excel file
- Geologic Time Scale information sheet

Your teacher will tell you how to access the Excel file.

Safety Precautions

Follow all normal lab safety rules.

Investigation Proposal Required?
☐ Yes   ☐ No

Getting Started

The Fossil Record 2 database is a near-complete listing of the diversity of life through geologic time, compiled at the level of the family (Benton 1993, 1995). In biology, levels of classification include species, genus, family, order, class, phylum, and kingdom (see Figure L5.2), so *family* refers to a level of classification that falls between order and genus. For example, *Pan paniscus*, a species of ape commonly called a bonobo, is a part of

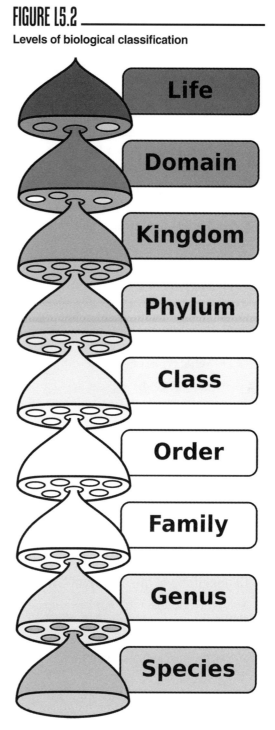

FIGURE L5.2

Levels of biological classification

Life

Domain

Kingdom

Phylum

Class

Order

Family

Genus

Species

the genus *Pan*, the family Hominidae, the order Primates, the class Mammalia, the phylum Chordata, and the kingdom Animalia.

The Fossil Record 2 database (Benton 1993) is an Excel file. Each row in this file represents a different biological family, and each column represents a different time interval. The geologic time range of each family is entered on the worksheet, with the first appearance of the family labeled as F1, presence in the fossil record labeled as 1, and last recorded appearance labeled as L1. You can download the entire database for free from this website: *http://palaeo.gly.bris.ac.uk/fossilrecord2/fossilrecord/index.html*.

You will not be using the entire database for this investigation; instead, you will be using an Excel file called Fossil Record 2 Database Summary Counts. This Excel file is a simplified version of The Fossil Record 2 database. It includes counts of the number of different families, orders, classes, or phyla found at specific time intervals in the geologic time scale. It also contains information about the number of families, orders, classes, or phyla that first appeared and were last observed in each time interval.

To answer the guiding question, you will need to analyze the data in the Fossil Record 2 Database Summary Counts. Be sure to think about the following questions before you begin analyzing your data:

- Are any of the data irrelevant based on the guiding question?
- How could you use mathematics to describe a change over time?
- What types of patterns might you look for as you analyze your data?
- How does the ratio of new families, orders, classes, or phyla and the total number of families, orders, classes, or phyla at each time interval compare with the other time intervals?
- How does the ratio of extinct families, orders, classes, or phyla and the total number of families, orders, classes, or phyla at each time interval compare with the other time intervals?
- Are there any other proportional relationships that you can look for that will help you answer the guiding question?
- What type of graph could you create to help make sense of your data?

Connections to the Nature of Scientific Knowledge and Scientific Inquiry

As you work through your investigation, be sure to think about

- how scientific knowledge changes over time, and
- how scientists use different methods to answer different types of questions.

Initial Argument

Once your group has finished collecting and analyzing your data, your group will need to develop an initial argument. Your initial argument needs to include a claim, evidence to support your claim, and a justification of the evidence. The *claim* is your group's answer to the guiding question. The *evidence* is an analysis and interpretation of your data. Finally, the *justification* of the evidence is why your group thinks the evidence matters. The justification of the evidence is important because scientists can use different kinds of evidence to support their claims. Your group will create your initial argument on a whiteboard. Your whiteboard should include all the information shown in Figure L5.3.

FIGURE L5.3 _____

Argument presentation on a whiteboard

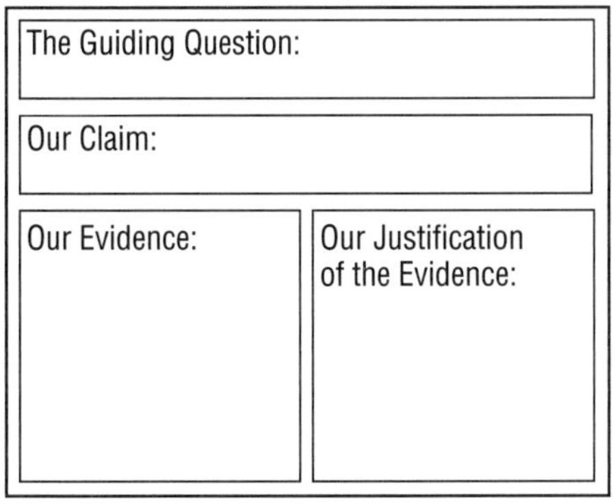

The Guiding Question:	
Our Claim:	
Our Evidence:	Our Justification of the Evidence:

Argumentation Session

The argumentation session allows all of the groups to share their arguments. One or two members of each group will stay at the lab station to share that group's argument, while the other members of the group go to the other lab stations to listen to and critique the other arguments. This is similar to what scientists do when they propose, support, evaluate, and refine new ideas during a poster session at a conference. If you are presenting your group's argument, your goal is to share your ideas and answer questions. You should also keep a record of the critiques and suggestions made by your classmates so you can use this feedback to make your initial argument stronger. You can keep track of specific critiques and suggestions for improvement that your classmates mention in the space below.

Critiques of our initial argument and suggestions for improvement:

If you are critiquing your classmates' arguments, your goal is to look for mistakes in their arguments and offer suggestions for improvement so these mistakes can be fixed. You should look for ways to make your initial argument stronger by looking for things that the other groups did well. You can keep track of interesting ideas that you see and hear during the argumentation in the space below. You can also use this space to keep track of any questions that you will need to discuss with your team.

Interesting ideas from other groups or questions to take back to my group:

Once the argumentation session is complete, you will have a chance to meet with your group and revise your initial argument. Your group might need to gather more data or design a way to test one or more alternative claims as part of this process. Remember, your goal at this stage of the investigation is to develop the best argument possible.

Report

Once you have completed your research, you will need to prepare an *investigation report* that consists of three sections. Each section should provide an answer for the following questions:

1. What question were you trying to answer and why?

2. What did you do to answer your question and why?

3. What is your argument?

Your report should answer these questions in two pages or less. You should write your report using a word processing application (such as Word, Pages, or Google Docs), if possible, to make it easier for you to edit and revise it later. You should embed any diagrams, figures, or tables into the document. Be sure to write in a persuasive style; you are trying to convince others that your claim is acceptable or valid.

References

Benton, M. J. 1993. *The fossil record 2.* London: Chapman & Hall.

Benton, M. J. 1995. Diversification and extinction in the history of life. Science 268 (5207): 52–58.

Gradstein, F. M., J. G. Ogg, and F. J. Hilgen. 2012. On the geologic time scale. *Newsletters on Stratigraphy* 45 (2): 171–188.

LAB 5

Lab 5. Geologic Time and the Fossil Record: Which Time Intervals in the Past 650 Million Years of Earth's History Are Associated With the Most Extinctions and Which Are Associated With the Most Diversification of Life?

Use the following diagram to answer questions 1–4. The picture shows a cross-section of Earth's strata and the fossils with the associated organisms discovered in those layers.

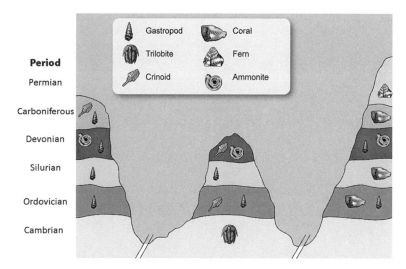

1. Crinoids evolved prior to ammonites.

    a. I agree with this statement

    b. I disagree with this statement

    How do you know?

Geologic Time and the Fossil Record

Which Time Intervals in the Past 650 Million Years of Earth's History Are Associated With the Most Extinctions and Which Are Associated With the Most Diversification of Life?

2. Which geologic period had the greatest diversification of life?

    a. Cambrian

    b. Ordovician

    c. Devonian

    How do you know?

3. Which geologic period had the most extinctions?

    a. Cambrian

    b. Ordovician

    c. Devonian

    How do you know?

4. The reason that crinoids do not appear in the Silurian period is because they went extinct and then re-evolved in the Devonian Period.

    a. I agree with this statement.

    b. I disagree with this statement.

How do you know?

5. Scientific knowledge does not change once it has been proven to be a fact.

    a. I agree with this statement.

    b. I disagree with this statement.

Explain your answer, using an example from your investigation about geologic time and the fossil record.

6. All scientific investigations follow the scientific method.

   a. I agree with this statement.

   b. I disagree with this statement.

   Explain your answer, using an example from your investigation about geologic time and the fossil record.

7. Scientists often need to look for patterns that occur in the data they collect and analyze. Explain why identifying patterns is important, using an example from your investigation about geologic time and the fossil record.

8. Natural phenomena occur at varying scales. Explain why scientists need to consider using different measurement or time scales when deciding how to collect and analyze data, using an example from your investigation about geologic time and the fossil record.

Plate Interactions

How Is the Nature of the Geologic Activity That Is Observed Near a Plate Boundary Related to the Type of Plate Interaction That Occurs at That Boundary?

Lab Handout

Lab 6. Plate Interactions: How Is the Nature of the Geologic Activity That Is Observed Near a Plate Boundary Related to the Type of Plate Interaction That Occurs at That Boundary?

Introduction

The interior structure of the Earth is composed of several layers (see Figure L6.1). At the center of the Earth is the inner core. The inner core is a solid sphere and consists of mostly iron. It has a radius of about 1,120 km. The next layer is the outer core. The outer core is liquid and extends beyond the inner core another 2,270 km. The next, and thickest, layer is the mantle. The mantle is often divided into three sublayers: the lower mesosphere, the upper mesosphere, and the asthenosphere. The outermost layer of the Earth is the lithosphere. The lithosphere includes the crust and the uppermost mantle.

The theory of plate tectonics states that the lithosphere is broken into several plates that move over time (see Figure L6.2). The plates move in different directions and at different speeds in relationship to each other. Plate boundaries are found where one plate interacts with another plate. These boundaries are classified into three different categories: (a) *convergent boundaries* result when two plates collide with each other, (b) *divergent boundaries* result when two plates move away from each other, and (c) *transform boundaries* form when two

FIGURE L6.1

Earth's layers

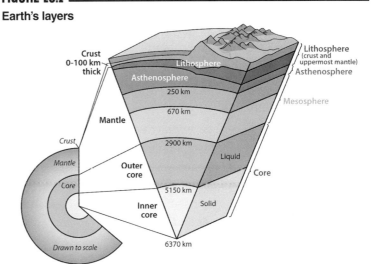

FIGURE L6.2

The major tectonic plates

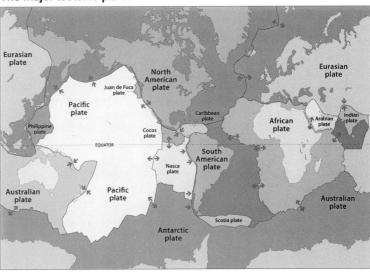

plates slide past each other. Volcanic eruptions and earthquakes often occur along or near plate boundaries.

In this investigation, you will explore where volcanic eruptions and earthquakes tend to happen. You goal is to determine if volcanic eruptions and earthquakes happen more often near a specific type of plate boundary. This type of investigation is important because natural processes, such as the gradual movement of tectonic plates over time, can result in natural hazards. Although it is impossible to prevent volcanic eruptions and earthquakes from happening, we can take steps to reduce their impacts. It is therefore useful for us to understand where these types of hazards are likely to occur because we can prepare for them and respond quickly when they happen. We can, for example, build better buildings, develop warning systems, and increase the response capabilities of cities to help reduce the loss of life and economic costs when we know where volcanic eruptions and earthquake tend to happen.

Your Task

Use an online interactive map to collect data about how often volcanic eruptions and earthquakes happen near the three different types of plate boundaries. Your goal is to use what you know about plate tectonics, patterns, and the use of different scales, proportional relationships, and quantities during an investigation to determine if the way plates interact with each other at a specific location is related to the occurrence of volcanic eruptions and earthquakes at that location.

The guiding question of this investigation is, *How is the nature of the geologic activity that is observed near a plate boundary related to the type of plate interaction that occurs at that boundary?*

Materials

You will use an online interactive map called *Natural Hazards Viewer* to conduct your investigation; the interactive map can be accessed at *http://maps.ngdc.noaa.gov/viewers/hazards*.

Safety Precautions

Be sure to follow all normal lab safety rules.

Investigation Proposal Required?   ☐ Yes   ☐ No

Getting Started

Given the nature of this investigation, you must determine what type of data you need to collect, how you will collect the data, and how will you analyze the data to answer the research question. To determine *what type of data you need to collect,* think about the following questions:

- How will you identify the location of different types of plate boundary?
- How can you describe an earthquake and a volcanic eruption quantitatively?
- What are the limitations of the available data set?

To determine *how you will collect the data*, think about the following questions:

- What parts of the world will you need to include in your study?
- What scale or scales should you use to quantify the size of an earthquake or a volcanic eruption?
- Will you need to limit the number of samples you include? If so, how will decide what to include?
- What concessions will you need to make to collect the data you need?
- How will you keep track of the data you collect and how will you organize it?

To determine *how you will analyze the data*, think about the following questions:

- What types of comparisons will you need to make?
- What types of patterns might you look for as you analyze the data?
- What potential proportional relationships can you find in the data?
- How could you use mathematics to determine if there are differences between the groups?
- What type of diagram could you create to help make sense of your data?

Connections to the Nature of Scientific Knowledge and Scientific Inquiry

As you work through your investigation, be sure to think about

- the difference between observations and inferences in science, and
- how the culture of science, societal needs, and current events influence the work of scientists.

Initial Argument

Once your group has finished collecting and analyzing your data, your group will need to develop an initial argument. Your initial argument needs to include a claim, evidence to support your claim, and a justification of the evidence. The *claim* is your group's answer to the guiding question. The *evidence* is an analysis and interpretation of your data. Finally, the *justification* of the evidence is why your group thinks the evidence matters. The justification of the evidence is important because scientists can use different kinds of evidence to support their claims. Your group will create your initial argument on a whiteboard. Your whiteboard should include all the information shown in Figure L6.3 (p. 76).

LAB 6

FIGURE L6.3 _____

Argument presentation on a whiteboard

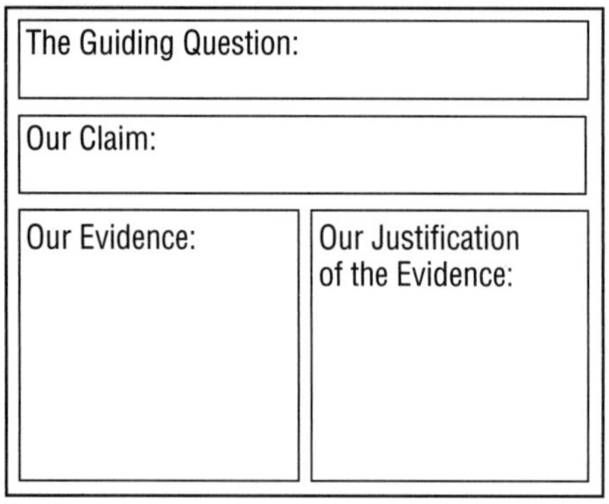

The Guiding Question:	
Our Claim:	
Our Evidence:	Our Justification of the Evidence:

Argumentation Session

The argumentation session allows all of the groups to share their arguments. One or two members of each group will stay at the lab station to share that group's argument, while the other members of the group go to the other lab stations to listen to and critique the other arguments. This is similar to what scientists do when they propose, support, evaluate, and refine new ideas during a poster session at a conference. If you are presenting your group's argument, your goal is to share your ideas and answer questions. You should also keep a record of the critiques and suggestions made by your classmates so you can use this feedback to make your initial argument stronger. You can keep track of specific critiques and suggestions for improvement that your classmates mention in the space below.

Critiques of our initial argument and suggestions for improvement:

If you are critiquing your classmates' arguments, your goal is to look for mistakes in their arguments and offer suggestions for improvement so these mistakes can be fixed. You should look for ways to make your initial argument stronger by looking for things that the other groups did well. You can keep track of interesting ideas that you see and hear during the argumentation in the space below. You can also use this space to keep track of any questions that you will need to discuss with your team.

Interesting ideas from other groups or questions to take back to my group:

Once the argumentation session is complete, you will have a chance to meet with your group and revise your initial argument. Your group might need to gather more data or design a way to test one or more alternative claims as part of this process. Remember, your goal at this stage of the investigation is to develop the best argument possible.

Report

Once you have completed your research, you will need to prepare an investigation report that consists of three sections. Each section should provide an answer for the following questions:

1. What question were you trying to answer and why?

2. What did you do to answer your question and why?

3. What is your argument?

Your report should answer these questions in two pages or less. You should write your report using a word processing application (such as Word, Pages, or Google Docs), if possible, to make it easier for you to edit and revise it later. You should embed any diagrams, figures, or tables into the document. Be sure to write in a persuasive style; you are trying to convince others that your claim is acceptable or valid.

Checkout Questions

Lab 6. Plate Interactions: How Is the Nature of the Geologic Activity That Is Observed Near a Plate Boundary Related to the Type of Plate Interaction That Occurs at That Boundary?

Use the map below to answer questions 1 and 2.

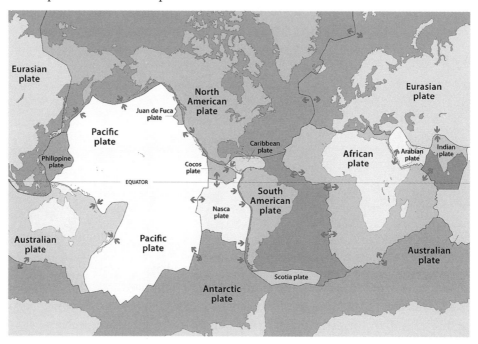

1. On the map above, circle one convergent boundary, one divergent boundary, and one transform boundary. Be sure to label each one. How do you know which boundary is which?

2. Earthquakes occur much more frequently in California than they do in Florida or New York. Using what you learned from your investigation and the map above, why is this the case?

Plate Interactions

How Is the Nature of the Geologic Activity That Is Observed Near a Plate Boundary Related to the Type of Plate Interaction That Occurs at That Boundary?

3. The map below shows the location of a volcanic arc in Central America. Each triangle represents the location of a different volcano.

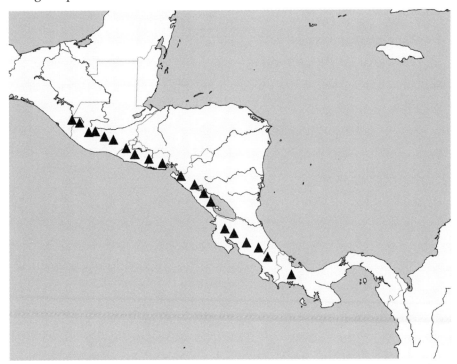

    a. What type of boundary is responsible for this volcanic arc and where is it most likely located?

    b. How do you know?

LAB 6

4. Scientists share a set of values, norms, and commitments that shape what counts as knowing, how to represent or communicate information, and how to interact with other scientists.

   a. I agree with this statement.

   b. I disagree with this statement.

   Explain your answer, using an example from your investigation about plate tectonics.

5. The statement "There were 31 earthquakes at the convergent boundary" is an example of an inference.

   a. I agree with this statement.

   b. I disagree with this statement.

   Explain your answer, using an example from your investigation about plate tectonics.

6. Scientists often need to look for patterns that occur in the data they collect and analyze. Explain why identifying patterns is important for scientists, using an example from your investigation about plate tectonics.

7. Natural phenomena occur at varying scales. Explain why scientists need to consider using different measurement or time scales when deciding how to collect and analyze data, using an example from your investigation about plate tectonics.

Application Lab

Lab Handout

Lab 7. Formation of Geologic Features: How Can We Explain the Growth of the Hawaiian Archipelago Over the Past 100 Million Years?

Introduction

Scientists use the theory of plate tectonics to explain current and past movements of the rocks at Earth's surface and the origin of many geologic features such as those shown in Figure L7.1. The theory of plate tectonics indicates that the lithosphere is broken into several plates that are in constant motion. Multiple lines of evidence support this theory. This evidence includes, but is not limited to, the location of earthquakes, chains of volcanoes (see Figure L7.1A), and non-volcanic mountain ranges (see Figure L7.1b) around the globe; how land under massive loads (such as lakes or ice sheets) can bend and even flow; the existence of mid-oceanic ridges; and the age of rocks near these ridges.

FIGURE L7.1

(a) The Aleutian archipelago, a chain of volcanic islands in Alaska; (b) the Himalayas, a nonvolcanic mountain range in Asia separating the plains of the Indian subcontinent from the Tibetan plateau

a   b

The plates are composed of oceanic and continental lithosphere. The plates move because they are located on top of giant convection cells in the mantle (see Figure L7.2). These currents bring matter from the hot inner mantle near the outer core up to the cooler surface and return cooler matter back to the inner mantle. The convection cells are driven by the energy that is released when isotopes deep within the interior of the Earth go through radioactive decay. The movement of matter in the mantle produces forces, which include viscous drag, slab pull, and ridge push, that together slowly move each of the plates across Earth's surface in a specific direction. The plates carry the continents, create or destroy ocean basins, form mountain ranges and plateaus, and produce earthquakes or volcanoes as they move.

Many interesting Earth surface features, such as the ones shown in Figure L7.1, are the result of either constructive or destructive geologic processes that occur along plate boundaries. There are three main types of plate boundaries (see Figure L7.3): *convergent boundaries* result when two plates collide with each other, *divergent boundaries* result when two plates move away from each other, and *transform boundaries* occur when plates slide past each other. We can explain many of the geologic features we see on Earth's surface when we understand how plates move and interact with each other over time.

Earth's surface is still being shaped and reshaped because of the movement of plates. One example of this phenomenon is the Hawaiian Islands. The Hawaiian Islands is an archipelago in the northern part of the Pacific Ocean that consists of eight major islands, several atolls, and numerous smaller islets. It extends from the island of Hawaii over 2,400 kilometers to the Kure Atoll. Each island is made up of one or more volcanoes (see Figure L7.4). The island of Hawaii, for instance, is made up of five different volcanoes. Two of the volcanoes found on the island of Hawaii are called Mauna Loa and Kilauea. Mauna Loa is the largest active volcano on Earth, and Kilauea is one of the most productive volcanoes in terms of how much lava erupts from it each year.

The number of islands in the Hawaiian archipelago has slowly increased over the last 100 million years. In this investigation, you will attempt to explain why these islands

FIGURE L7.2 _____

Convection cells in the mantle

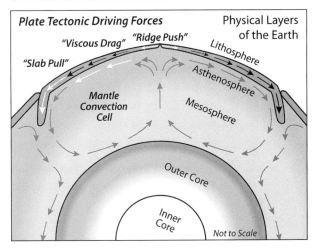

FIGURE L7.3 _____

The three types of plate boundaries

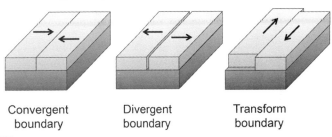

Convergent boundary   Divergent boundary   Transform boundary

FIGURE L7.4 _____

The Hawaiian archipelago and some of its volcanoes

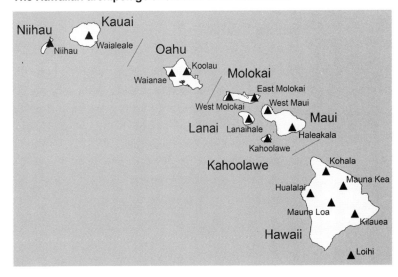

are in the middle of the Pacific Ocean, why they form a chain instead of some other shape, why some of the islands are bigger than other ones, and why the number of islands in the archipelago has slowly increased over time.

Your Task

Develop a conceptual model that you can use to explain how the Hawaiian archipelago formed over the last 100 million years. Your conceptual model must be based on what we know about patterns, systems and system models, and the movement of Earth's plates over time. You should be able to use your conceptual model to predict when and where you will see a new island appear in the Hawaiian archipelago.

The guiding question of this investigation is, *How can we explain the growth of the Hawaiian archipelago over the past 100 million years?*

Materials

You will use a computer with Excel or other spreadsheet application during your investigation. You will also use the following resources:

- *Natural Hazards Viewer* online interactive map, available at *http://maps.ngdc.noaa. gov/viewers/hazards*
- Ages of Volcanoes in Hawaiian Islands Excel file; your teacher will tell you how to access the Excel file.

Safety Precautions

Follow all normal lab safety rules.

Investigation Proposal Required?   ☐ Yes   ☐ No

Getting Started

The first step in the development of your conceptual model is to learn as much as you can about the geologic activity around the Hawaiian archipelago. You can use the *Natural Hazards Viewer* interactive map to determine the location of any plate boundaries around the islands, the location of volcanoes on and around each island, and the occurrence and magnitude of earthquakes in the area. As you use the *Natural Hazards Viewer*, be sure to consider the following questions:

- What are the boundaries and the components of the system you are studying?
- How do the components of this system interact with each other?
- How can you quantitatively describe changes within the system over time?
- What scale or scales should you use to when you take your measurements?

- What is going on at the unobservable level that could cause the things that you observe?

The second step in the development of your conceptual model is to learn more about the characteristics of the volcanoes in the Hawaiian archipelago. You can use the Excel file called Ages of Volcanoes in Hawaiian Islands to determine which volcanoes are active and which are dormant, the distances between the volcanoes, and the age of each volcano. As you analyze the data in this Excel file, be sure to consider the following questions:

- What types of patterns could you look for in your data?
- How could you use mathematics to describe a relationship between two variables?
- What could be causing the pattern that you observe?
- What graphs could you create in Excel to help you make sense of the data?

Once you have learned as much as you can about Hawaiian archipelago system, your group can begin to develop your conceptual model. A conceptual model is an idea or set of ideas that explains what causes a particular phenomenon in nature. People often use words, images, and arrows to describe a conceptual model. Your conceptual model needs to be able to explain the origin of the Hawaiian archipelago. It also needs to be able to explain

- why the islands form a chain and not some other shape,
- why the number of islands has increased over the last 100 million years,
- why some islands are bigger than other ones, and
- what will likely happen to the Hawaiian archipelago over the next 100 million years.

The last step in your investigation will be to generate the evidence you to need to convince others that your model is valid and acceptable. To accomplish this goal, you can attempt to show how using a different version of your model or making a specific change to a portion of your model would make your model inconsistent with what we know about the islands in the Hawaiian archipelago. Scientists often make comparisons between different versions of a model in this manner to show that a model they have developed is valid or acceptable. You can also use the *Natural Hazards Viewer* to identify other chains of volcanoes that are similar to ones found in the Hawaiian archipelago. You can then determine if you are able to use your model to explain the formation of other chains of volcanoes. If you are able to show how your conceptual model explains the formation of the Hawaiian archipelago better than other models or that you can use your conceptual model to explain many different phenomena, then you should be able to convince others that it is valid or acceptable.

LAB 7

Connections to the Nature of Scientific Knowledge and Scientific Inquiry

As you work through your investigation, be sure to think about

- the use of models as tools for reasoning about natural phenomena in science, and
- the assumptions made by scientists about order and consistency in nature.

Initial Argument

Once your group has finished collecting and analyzing your data, your group will need to develop an initial argument. Your initial argument needs to include a claim, evidence to support your claim, and a justification of the evidence. The *claim* is your group's answer to the guiding question. The *evidence* is an analysis and interpretation of your data. Finally, the

FIGURE L7.5 _____
Argument presentation on a whiteboard

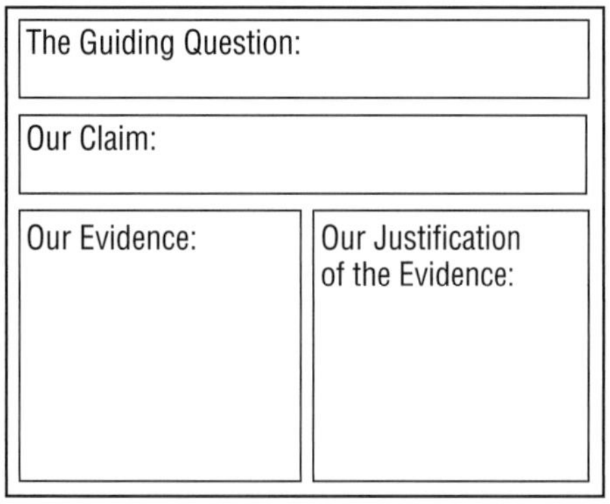

The Guiding Question:	
Our Claim:	
Our Evidence:	Our Justification of the Evidence:

justification of the evidence is why your group thinks the evidence matters. The justification of the evidence is important because scientists can use different kinds of evidence to support their claims. Your group will create your initial argument on a whiteboard. Your whiteboard should include all the information shown in Figure L7.5.

Argumentation Session

The argumentation session allows all of the groups to share their arguments. One or two members of each group will stay at the lab station to share that group's argument, while the other members of the group go to the other lab stations to listen to and critique the other arguments. This is similar to what scientists do when they propose, support, evaluate, and refine new ideas during a poster session at a conference. If you are presenting your group's argument, your goal is to share your ideas and answer questions. You should also keep a record of the critiques and suggestions made by your classmates so you can use this feedback to make your initial argument stronger. You can keep track of specific critiques and suggestions for improvement that your classmates mention in the space below.

Critiques of our initial argument and suggestions for improvement:

If you are critiquing your classmates' arguments, your goal is to look for mistakes in their arguments and offer suggestions for improvement so these mistakes can be fixed. You should look for ways to make your initial argument stronger by looking for things that the other groups did well. You can keep track of interesting ideas that you see and hear during the argumentation in the space below. You can also use this space to keep track of any questions that you will need to discuss with your team.

Interesting ideas from other groups or questions to take back to my group:

Once the argumentation session is complete, you will have a chance to meet with your group and revise your initial argument. Your group might need to gather more data or design a way to test one or more alternative claims as part of this process. Remember, your goal at this stage of the investigation is to develop the best argument possible.

Report

Once you have completed your research, you will need to prepare an *investigation report* that consists of three sections. Each section should provide an answer for the following questions:

1. What question were you trying to answer and why?

2. What did you do to answer your question and why?

3. What is your argument?

Your report should answer these questions in two pages or less. You should write your report using a word processing application (such as Word, Pages, or Google Docs), if possible, to make it easier for you to edit and revise it later. You should embed any diagrams, figures, or tables into the document. Be sure to write in a persuasive style; you are trying to convince others that your claim is acceptable or valid.

LAB 7

Lab 7. Formation of Geologic Features: How Can We Explain the Growth of the Hawaiian Archipelago Over the Past 100 Million Years?

1. Below is a map of the Hawaiian archipelago, shown from above. On the map, draw what you think the archipelago will look like in 100 million years.

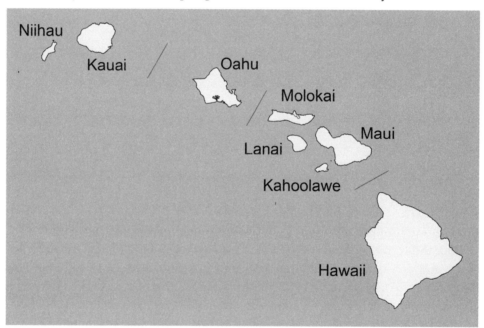

Explain your drawing below.

2. Below is a picture of the Japanese archipelago. What information would you need to determine if the Japanese archipelago formed in the same way the Hawaiian archipelago formed?

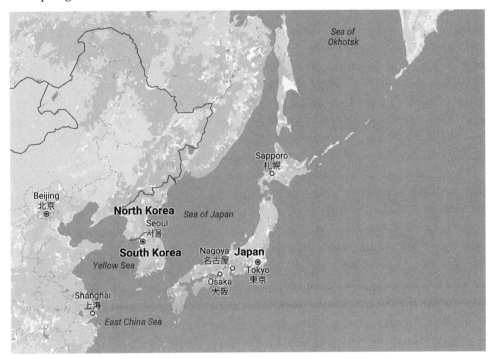

3. Scientists can change or refine a model when presented with new evidence.

   a. I agree with this statement.

   b. I disagree with this statement.

Explain your answer, using an example from your investigation about the formation of the Hawaiian archipelago.

4. When trying to understand events that happened in the past, scientists assume that natural laws operate today in the same way as they did in the past.

   a. I agree with this statement.

   b. I disagree with this statement.

   Explain your answer, using an example from your investigation about the formation of the Hawaiian archipelago.

5. In science, it is important to define a system under study and then develop a model of the system. Explain why this is important to do, using an example from your investigation about the formation of the Hawaiian archipelago.

6. Scientists often look for patterns as part of their work. Explain why it is important to identify patterns during an investigation, using an example from your investigation about the formation of the Hawaiian archipelago.

SECTION 4
Earth's Systems

Introduction Labs

Lab Handout

Lab 8. Surface Erosion by Wind: Why Do Changes in Wind Speed, Wind Duration, and Soil Moisture Affect the Amount of Soil That Will Be Lost Due to Wind Erosion?

Introduction

Earth scientists use the term *erosion* to describe what happens when liquid water, ice, or wind moves rock, soil, or dissolved materials from one location to another. The process of erosion is responsible for the creation of many interesting landforms or natural phenomena that we observe in the world around us. The rock shown in Figure L8.1, for example, was sculpted by wind erosion. Wind erosion can also produce dust storms such as the one that moved through Casa Grande, Arizona, on July 5, 2011. This enormous storm was 101 miles wide and 7,000 feet high, with wind gusts reaching 60 miles per hour (mph). It obscured mountains and covered freeways and homes with massive amounts of soil (see Figure L8.2).

FIGURE L8.1 _____

A rock sculpted by wind erosion in the Altiplano region of Bolivia

FIGURE L8.2 _____

A dust storm in Casa Grande, Arizona

Wind is the movement of air. It is caused by differences in air pressure within the atmosphere. Wind erodes the surface of Earth through abrasion or deflation. *Abrasion* is the wearing down of rock or other objects. Wind abrasion over a long period of time can make significant changes to the shape of a rock. In fact, the rock in Figure L8.1 looks the way that it does because of wind abrasion. *Deflation* is the removal of soil particles, such as sand, silt, and clay, from an area by wind. Wind can remove soil particles from one location and move them to a different location because moving air exerts a pushing force on objects. Wind can move soil particles by suspending them in the air or causing them to bounce or slide along the ground. Soil particles will continue to move until the pushing force of the wind acting on them weakens or stops. These soil particles often accumulate and form mounds or dunes around physical barriers. The dust storm in Figure L8.1 is an example of wind deflation. This dust storm moved a massive amount of soil into the city of Casa Grande.

Wind erosion is a big problem in some regions of the United States because wind can remove and transport a lot of soil in these regions. Some regions struggle with the effects of wind erosion and some regions do not because different regions of the United States have different wind patterns, types of soil, and climates. In some regions of the United States, for example, the average wind speed never exceeds 4.0 mph, but in other areas the average is consistently above 10.5 mph. Wind speed is important to consider because fast-moving air applies a greater pushing force than slow-moving air. The soil found in a region differs in terms of the proportion of sand, silt, and clay that is present in it.

Earth scientists use the soil classification triangle shown in Figure L8.3 to classify soils. Figure L8.4 shows the different types of soil particles and their relative sizes. The size of soil particles is important to consider because objects with more mass require more force to move. The climate of a region affects how much rain a region will typically have at different times of the year. The amount of rainfall affects the moisture of the soil.

Wind speed, wind duration, and soil moisture, as well as several other factors (such as amount and type of vegetation in a region), will affect the amount of soil that will be lost from a given area due to wind erosion. Soil loss due to wind erosion

FIGURE L8.3
Soil types by composition

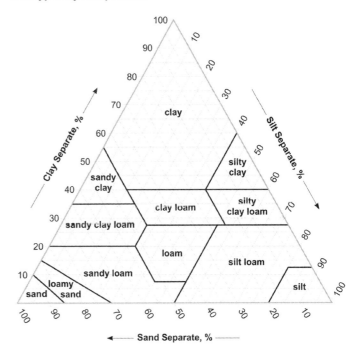

FIGURE L8.4
Soil particle types and sizes

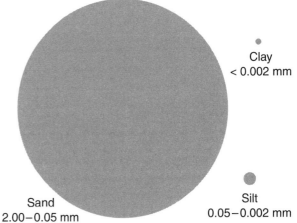

can destroy crops, put farmers out of business, lead to food shortages, and devastate the economy. During the 1930s, for example, winds removed so much soil on farms in the Midwest that entire crops failed. The winds that blew across the plains suspended so much topsoil in the air that it sometimes blackened the sky. These billows of dust, which were called black blizzards or black rollers at the time, transported soil across the country and even deposited some of it in East Coast cities such as Washington, D.C., and New York. Many farmers went out of business, and the United States, as a result, faced massive food shortages. To prevent a crisis like this from happening again, people must understand how and why soil erodes due to wind. You will therefore explore how wind speed, wind duration, and soil moisture affect the amount of soil that is lost to wind erosion and then develop a conceptual model that you can use to explain your observations and predict how much soil will be lost due to wind erosion under different conditions.

Your Task

Use what you know about wind erosion, soil types, factors that affect rates of change, and the importance of quantity, scale, and proportion in science to determine *how* wind speed, wind duration, and soil moisture affect the amount of soil lost from different types of soil. Then develop a conceptual model that can be used to explain *why* these factors affect the amount of soil lost due to wind erosion. Once you have developed your conceptual model, you will need to test it using a different type of soil to determine if it allows you to predict how much soil will be lost due to wind erosion under a different condition.

The guiding question of this investigation is, *Why do changes in wind speed, wind duration, and soil moisture affect the amount of soil that will lost due to wind erosion?*

Materials

You may use any of the following materials during your investigation:

Consumables	Equipment	
• Soil sample A	• Safety glasses or goggles (required)	• 2 ml Beakers (each 250 ml)
• Soil sample B	• Chemical-resistant apron (required)	• Graduated cylinder (100 ml)
• Soil sample C	• Gloves (required)	• Fan with multiple speeds
• Soil sample D	• Cookie sheet	• Electronic or triple beam balance
• Water	• Trash bag	• Flowchart for soil texture by feel

Safety Precautions

Follow all normal lab safety rules. In addition, take the following safety precautions:

- Wear sanitized indirectly vented chemical-splash goggles and chemical-resistant, nonlatex aprons and gloves throughout the entire investigation (which includes setup and cleanup).

Surface Erosion by Wind

Why Do Changes in Wind Speed, Wind Duration, and Soil Moisture Affect the Amount of Soil That Will Be Lost Due to Wind Erosion?

- Handle all glassware with care.

- Plug the fan only into a GFCI-protected circuit. Keep the fan away from water to avoid risk of shock.

- Use caution when working with the fan. Blades can be sharp and cut skin. Do not place fingers into the fan, and never place any object into the turning fan blade.

- Report and clean up any spills immediately, and avoid walking in areas where water has been spilled.

- Wash hands with soap and water when done collecting the data and after completing the lab.

Investigation Proposal Required?   ☐ Yes   ☐ No

Getting Started

The first step in developing your model is to plan and carry out three different experiments to determine how wind speed, wind duration, and soil moisture affect the amount of soil lost from different types of soil. You will use soil samples A, B, and C during your experiments. To design each experiment, you will need to determine what type of data you need to collect, how you will collect it, and how you will analyze it.

To determine *what type of data you need to collect*, think about the following questions:

- How will you identify the type of soil?

- How will you determine wind speed?

- How will you determine soil moisture quantitatively?

- How will you determine the amount of soil lost due to wind erosion quantitatively?

- How can you describe the other components of the system?

- How could you keep track of changes over time quantitatively?

To determine *how you will collect the data*, think about the following questions:

- What will be the independent variable and the dependent variable in each experiment?

- What will be the treatment and control conditions in each experiment?

- What other factors will you need to keep constant?

- What scale or scales should you use when you take your measurements?

- What equipment will you need to collect the data you need?

- How will you make sure that your data are of high quality (i.e., how will you reduce error)?

- How will you keep track of and organize the data you collect?

To determine *how you will analyze the data,* think about the following questions:

- How could you use mathematics to describe a change over time?
- How could you use mathematics to determine if there is a difference between the experimental conditions or a relationship between variables?
- What types of proportional relationships might you look for as you analyze your data?
- What type of calculations will you need to make?
- What type of table or graph could you create to help identify a trend in the data?

Once you have carried out your three experiments and understand how changes in wind speed, wind duration, and soil moisture affect the amount of soil lost from different types of soil, your group will need to develop a conceptual model. The model needs to be able to explain why these three factors affect the amount of soil lost to wind erosion in the way that they do. The model also needs to account for the physical properties of the particles that make up soil.

The last step in this investigation is to test your model. To accomplish this goal, you can use soil sample D to determine if your model leads to accurate predictions about the amount of soil that will be lost from a different type of soil under specific conditions (e.g., high wind speed, short duration, and low moisture). If you are able to use your model to make accurate predictions about the effects of wind erosion under different conditions, then you will be able to generate the evidence you need to convince others that the conceptual model you developed is valid or acceptable.

Connections to the Nature of Scientific Knowledge and Scientific Inquiry

As you work through your investigation, be sure to think about

- the difference between data and evidence in science, and
- the nature and role of experiments in science.

Initial Argument

Once your group has finished collecting and analyzing your data, your group will need to develop an initial argument. Your initial argument needs to include a claim, evidence to support your claim, and a justification of the evidence. The *claim* is your group's answer to the guiding question. The *evidence* is an analysis and interpretation of your data. Finally, the *justification* of the evidence is why your group thinks the evidence matters. The justification of the evidence is important because scientists can use different kinds of evidence to support their claims. Your group will create your initial argument on a whiteboard. Your whiteboard should include all the information shown in Figure L8.5.

Argumentation Session

The argumentation session allows all of the groups to share their arguments. One or two members of each group will stay at the lab station to share that group's argument, while the other members of the group go to the other lab stations to listen to and critique the other arguments. This is similar to what scientists do when they propose, support, evaluate, and refine new ideas during a poster session at a conference. If you are presenting your group's argument, your goal is to share your ideas and answer questions. You should also keep a record of the critiques and suggestions made by your classmates so you can use this feedback to make your initial argument stronger. You can keep track of specific critiques and suggestions for improvement that your classmates mention in the space below.

FIGURE L8.5

Argument presentation on a whiteboard

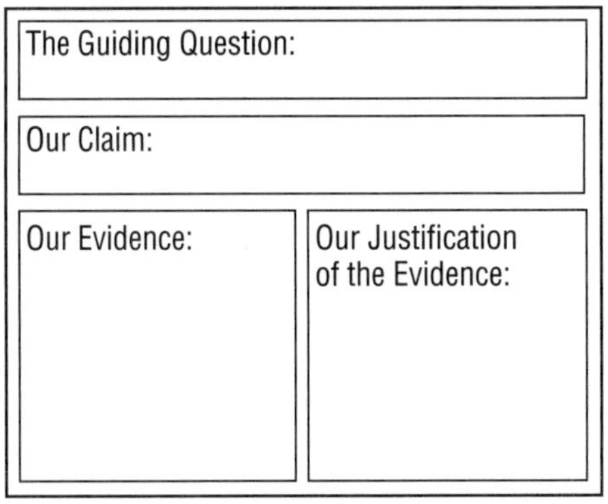

The Guiding Question:	
Our Claim:	
Our Evidence:	Our Justification of the Evidence:

Critiques of our initial argument and suggestions for improvement:

If you are critiquing your classmates' arguments, your goal is to look for mistakes in their arguments and offer suggestions for improvement so these mistakes can be fixed. You should look for ways to make your initial argument stronger by looking for things that the other groups did well. You can keep track of interesting ideas that you see and hear during the argumentation in the space below. You can also use this space to keep track of any questions that you will need to discuss with your team.

Interesting ideas from other groups or questions to take back to my group:

Once the argumentation session is complete, you will have a chance to meet with your group and revise your initial argument. Your group might need to gather more data or design a way to test one or more alternative claims as part of this process. Remember, your goal at this stage of the investigation is to develop the best argument possible.

Report

Once you have completed your research, you will need to prepare an *investigation* report that consists of three sections. Each section should provide an answer for the following questions:

1. What question were you trying to answer and why?

2. What did you do to answer your question and why?

3. What is your argument?

Your report should answer these questions in two pages or less. You should write your report using a word processing application (such as Word, Pages, or Google Docs), if possible, to make it easier for you to edit and revise it later. You should embed any diagrams, figures, or tables into the document. Be sure to write in a persuasive style; you are trying to convince others that your claim is acceptable or valid.

Checkout Questions

Lab 8. Surface Erosion by Wind: Why Do Changes in Wind Speed, Wind Duration, and Soil Moisture Affect the Amount of Soil That Will Be Lost Due to Wind Erosion?

1. A scientist has a collection of soil samples, each classified as shown in the table below.

Sample	Sample classification	Soil moisture (% water by volume)
A	Loamy sand	19
B	Silty clay loam	25
C	Silt loam	10
D	Silty clay loam	15

   a. If exposed to the same wind speed for an equal duration of time, which soil will likely lose the greatest mass to wind erosion?

   b. How do you know? Explain using an example from your investigation about the factors affecting soil loss due to wind erosion.

c. If exposed to the same wind speed for an equal duration of time, which soil will likely lose the least mass to wind erosion?

d. How do you know? Explain using an example from your investigation about the factors affecting soil loss due to wind erosion.

2. Given what you know about surface erosion and using examples from your investigation, what are some steps that scientists and engineers can take to help prevent soil erosion in the following areas?

   a. A mountain range

Surface Erosion by Wind

Why Do Changes in Wind Speed, Wind Duration, and Soil Moisture Affect the Amount of Soil That Will Be Lost Due to Wind Erosion?

    b. A park in your local community

    c. Near the ocean or a lake

3. Scientists use experiments to test the validity of a hypothesis (i.e., a tentative explanation) for an observed phenomenon.

    a. I agree with this statement.

    b. I disagree with this statement.

Explain your answer, using an example from your investigation about the factors affecting soil loss due to wind erosion.

LAB 8

4. In science, there is a difference between data and evidence.

    a. I agree with this statement.

    b. I disagree with this statement.

    Explain your answer, using an example from your investigation about the factors affecting soil loss due to wind erosion.

5. In nature, events can occur at varying scales of size and of time. Why is it important to consider the differences in size and scale during an investigation? Give an example of a relatively small effect of wind erosion and a relatively large effect of wind erosion.

Surface Erosion by Wind

Why Do Changes in Wind Speed, Wind Duration, and Soil Moisture Affect the Amount of Soil That Will Be Lost Due to Wind Erosion?

6. In science, it is important to understand what factors influence rates of change in system. Explain why this is so important, using an example from your investigation about the factors affecting soil loss due to wind erosion.

Lab 9. Sediment Transport by Water: How Do Changes in Stream Flow Affect the Size and Shape of a River Delta?

Introduction

A *drainage basin* is all the land that contributes water to a river system. The land within a drainage basin is marked on a map by watershed boundaries. A *watershed boundary* is the location where water will either flow into a given drainage basin or flow into a different one. Watershed boundaries follow ridgelines because water always flows from a higher elevation to a lower elevation. Figure L9.1 is a map of the Mississippi River drainage basin, which is the largest drainage basin in North America. It extends between the Rocky Mountains in the west and the Appalachian Mountains in the east. The Mississippi River and its tributaries collect water from more than 3.2 million square kilometers (1.2 million square miles) of the continent. All this water empties into the Gulf of Mexico in Louisiana.

A *river* is water that flows in a channel. Water makes its way to the sea in a channel under the influence of gravity. The time required for the journey depends on the velocity of the river. *Velocity* is the distance the water travels in a unit of time. The ability of a river to erode and transport materials depends on its velocity. Even slight changes in velocity can lead to significant changes in the load of sediment that water can transport. Several factors

FIGURE L9.1

The drainage basin of the Mississippi River

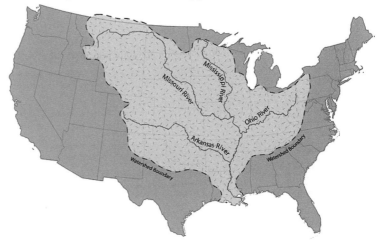

affect the velocity of a river. These factors include the gradient of the channel; the shape, size, and roughness of the channel; and the stream flow.

The *gradient* of a river channel is the vertical drop of a river over a specified distance. The velocity of a river increases as the gradient of a channel increases. The shape, size, and roughness of the channel, however, also affect the velocity of river because water encounters friction as it moves through a channel. Water moves faster through straight, large, and smooth channels than it does through irregular, small, and rough ones. The *stream flow* of a river is the volume of water flowing past a certain point in a given unit of time. The stream flow of most rivers is not constant because it is influenced by the amount of surface water runoff that is currently in the drainage basin. In general, the more surface water runoff that is in a drainage basin, the greater the stream flow of a river. Surface water runoff comes from rain or melting snow.

Rivers create different types of landforms because of the way they are able to produce, transfer, and deposit large amounts of sediment. River deltas are an example of a landform that is produced by a river. River deltas form where a river meets a larger body of water, such as a lake, gulf, or ocean. When the two bodies of water meet, the water from the river cannot flow at the same speed it did in the river channel. As the stream flow slows, the river is no longer ability to carry sediment. The sediment therefore accumulates in large amounts where the two bodies of water meet. Over a long period of time, this sediment builds into a fanlike structure such as the delta of the Mississippi River, seen in Figure L9.2.

FIGURE L9.2

The Mississippi River delta; notice the amount of sediment deposited where the river enters the Gulf of Mexico (marked with a circle)

Native vegetation flourishes around river deltas because the soil is rich in minerals and organic nutrients. The land around a river delta is also excellent for growing crops. River deltas also create wetlands and marshes that protect the shorelines and *estuaries* (transition zones between river and marine environments) because they absorb excess runoff from both floods and storms. River deltas provide unique habitats for many different species that live along the coast and within the estuaries. Many species, as a result, need these habitats in order to grow and thrive.

Humans often change the characteristics of rivers to better suit their needs. For example, humans build dams to generate electricity, create levees to protect buildings or homes from flooding, and dig channels to divert water from rivers to farms in order to produce crops. These are just a few examples of the many ways humans alter the characteristics of rivers. The addition of a dam, levee, or channel anywhere in a drainage basin will change the stream flow of a river. It will also affect the amount and type of sediment in the water that is able to reach the mouth of a river. Human activity that changes the characteristics of a river, as a result, will affect how water moves through a drainage basin and could contribute to a number of environmental, economic, or social problems. In this investigation, you have an opportunity to learn how changing the stream flow of a river affects the size or shape of its delta.

LAB 9

Your Task

Use a stream table to create a physical model of a river that empties into a larger body of water. Then use this physical model and what you know about erosion, sediments, how matter flows within a system, and scales, proportion, and quantity to design and carry out an investigation in order to determine the relationship between the river stream flow and the size or shape of its river delta.

The guiding question of this investigation is, *How do changes in stream flow affect the size and shape of a river delta?*

Materials

You may use any of the following materials during your investigation:

Consumables	Equipment	
• Water for the stream table	• Safety glasses or goggles (required)	• Graduated cylinder (250 ml)
• Sand (for the stream table)	• Chemical-resistant apron (required)	• Electronic or triple beam balance
• Paper towels	• Gloves (required)	• Stopwatch
	• Stream table with water volume adjustment	• Funnel
	• Plastic sheet (12" × 18")	• Ruler
	• 1–2 Beakers (each 1,000 ml)	• Protractor

Safety Precautions

Follow all normal lab safety rules. Be sure to use the stream table as instructed by your teacher. In addition, take the following safety precautions:

- Wear sanitized indirectly vented chemical-splash goggles and chemical-resistant, nonlatex aprons and gloves throughout the entire investigation (which includes setup and cleanup).

- Keep away from electrical outlets when working with water to prevent or reduce risk of shock.

- Report and clean up spills immediately, and avoid walking in areas where water has been spilled.

- Handle all glassware with care.

- Wash hands with soap and water when done collecting the data and after completing the lab.

Investigation Proposal Required?   ☐ Yes   ☐ No

Getting Started

To answer the guiding question, you will need to carry out an experiment using a stream table. The stream table simulates the behavior of a drainage basin. You can set up the stream table as shown in Figure L9.3. When you add sand, carve a channel into the sand, and then add water to one end of the channel, a river will flow through the channel. The delta will then form on the plastic sheet.

Before you collect your data, spend some time familiarizing yourself with the stream table that you will use. Once you understand how the stream table works, you can plan your experiment. To accomplish this task, you must determine what type of data you need to collect, how you will collect it, and how will you analyze it.

To determine *what type of data you need to collect*, think about the following questions:

- What are the boundaries and components of the system you are studying?
- How can you describe the components of the system quantitatively?
- How can you track how matter flows into, out of, or within this system?
- How will you quantify the size and shape of the delta?

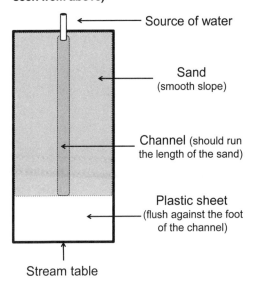

FIGURE L9.3

How to set up a stream table (image is as seen from above)

- Source of water
- Sand (smooth slope)
- Channel (should run the length of the sand)
- Plastic sheet (flush against the foot of the channel)
- Stream table

To determine *how you will collect the data*, think about the following questions:

- What will be your independent variable and dependent variables in the experiment?
- What will your treatment conditions be and how will you set them up?
- What will serve as a control condition?
- How many trials will be run for each condition?
- How long will each trial run before data are collected?
- When will you need to take measurements?
- What measurement scale or scales should you use to collect data?
- How will you make sure that your data are of high quality (i.e., how will you reduce error)?

- How will you keep track of and organize the data you collect?

To determine *how you will analyze the data,* think about the following questions:

- How could you use mathematics to describe a difference between conditions or a relationship between variables?
- What types of patterns might you look for as you analyze your data?
- What type of calculations will you need to make?
- What type of table or graph could you create to help make sense of your data?

Connections to the Nature of Scientific Knowledge and Scientific Inquiry

As you work through your investigation, be sure to think about

- the use of models as tools for reasoning about natural phenomena, and
- the nature and role of experiments in science.

Initial Argument

Once your group has finished collecting and analyzing your data, your group will need to develop an initial argument. Your initial argument needs to include a claim, evidence to support your claim, and a justification of the evidence. The claim is your group's answer to the guiding question. The evidence is an analysis and interpretation of your data. Finally, the justification of the evidence is why your group thinks the evidence matters. The justification of the evidence is important because scientists can use different kinds of evidence to support their claims. Your group will create your initial argument on a whiteboard. Your whiteboard should include all the information shown in Figure L9.4.

FIGURE L9.4 ⎯⎯⎯⎯⎯⎯⎯⎯

Argument presentation on a whiteboard

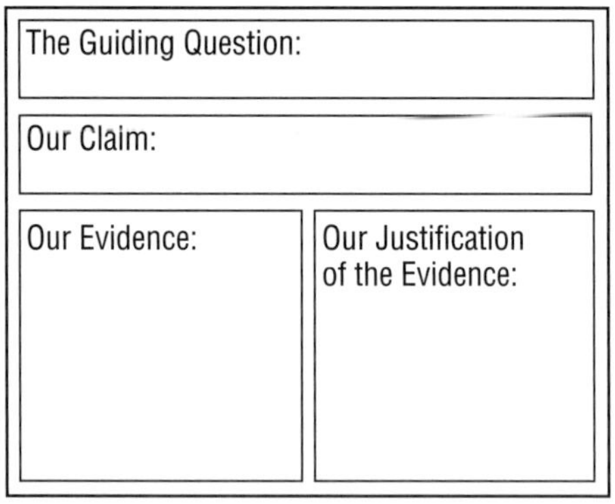

The Guiding Question:	
Our Claim:	
Our Evidence:	Our Justification of the Evidence:

Argumentation Session

The argumentation session allows all of the groups to share their arguments. One or two members of each group will stay at the lab station to share that group's argument, while the other members of the group go to the other lab stations to listen to and critique the other arguments. This is similar to what scientists do when they propose, support, evaluate, and refine new ideas during a poster session at a conference. If you are presenting your group's argument, your goal is to share your ideas and answer questions. You should also keep a record of the critiques and suggestions made by your classmates so you can use this feedback to make your initial argument stronger. You can keep track of specific critiques and suggestions for improvement that your classmates mention in the space below.

Critiques of our initial argument and suggestions for improvement:

If you are critiquing your classmates' arguments, your goal is to look for mistakes in their arguments and offer suggestions for improvement so these mistakes can be fixed. You should look for ways to make your initial argument stronger by looking for things that the other groups did well. You can keep track of interesting ideas that you see and hear during the argumentation in the space below. You can also use this space to keep track of any questions that you will need to discuss with your team.

LAB 9

Interesting ideas from other groups or questions to take back to my group:

Once the argumentation session is complete, you will have a chance to meet with your group and revise your initial argument. Your group might need to gather more data or design a way to test one or more alternative claims as part of this process. Remember, your goal at this stage of the investigation is to develop the best argument possible.

Report

Once you have completed your research, you will need to prepare an investigation report that consists of three sections. Each section should provide an answer for the following questions:

1. What question were you trying to answer and why?

2. What did you do to answer your question and why?

3. What is your argument?

Your report should answer these questions in two pages or less. You should write your report using a word processing application (such as Word, Pages, or Google Docs), if possible, to make it easier for you to edit and revise it later. You should embed any diagrams, figures, or tables into the document. Be sure to write in a persuasive style; you are trying to convince others that your claim is acceptable or valid.

Checkout Questions

Lab 9. Sediment Transport by Water: How Do Changes in Stream Flow Affect the Size and Shape of a River Delta?

Use the figure below to answer questions 1 and 2.

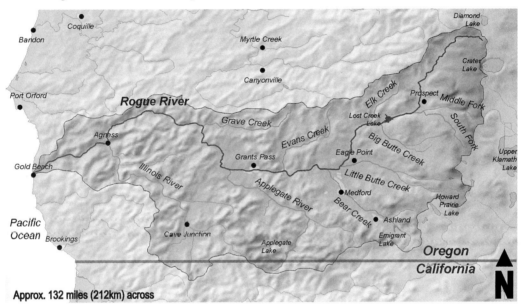

1. A dam is constructed near the town of Agness. What effect will this likely have on the amount of sediment that reaches the town of Gold Beach? Explain why the dam could cause these effects.

2. There is more snowmelt than normal. What effect will this likely have on the amount of sediment that reaches the town of Grants Pass? Explain why an increase in snowmelt could cause these effects.

3. Scientists use experiments to prove a hypothesis is correct.

    a. I agree with this statement.

    b. I disagree with this statement.

    Explain your answer, using an example from your investigation about sediment transport by water.

4. A model is a three-dimensional representation of something on a smaller scale than the original.

   a. I agree with this statement.

   b. I disagree with this statement.

   Explain your answer, using an example from your investigation about sediment transport by water.

5. Scientists often need to be able to track how energy and matter move into, out of, and within systems during an investigation. Explain why it is important track energy and matter, using an example from your investigation about sediment transport by water.

6. Scientists often need to consider what measurement scale or scales to use during an investigation. Explain why it is important for scientist to think about the measurement scales, using an example from your investigation about sediment transport by water.

Lab Handout

Lab 10. Deposition of Sediments: How Can We Explain the Deposition of Sediments in Water?

Introduction

People use sedimentary rocks such as siltstone, shale, and sandstone (see Figure L10.1) for many different purposes. Siltstone, for example, is used to build homes and walls or for decorations. Shale is used to make cement, terra-cotta pots, bricks, and roof tiles. Sandstone is another sedimentary rock that is used as a building material. It is used to create floor tiles and decorative walls in homes or businesses and to create monuments and roads. Sandstone is also used as a sharpening stone for knives.

FIGURE L10.1

Some examples of sedimentary rock

Siltstone

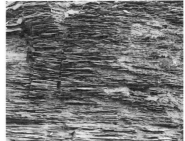
Shale

Sandstone

The material that makes up siltstone, shale, and sandstone comes from other rocks that have weathered over time. These rocks are therefore called *clastic sedimentary rocks*. As a rock weathers, it is broken into smaller pieces or sediments. These sediments are then carried to other places by wind, liquid water, or glacial ice. The sediments eventually settle out of the air or water and accumulate at a specific location. This process is called *deposition,* and it results in layers of different types of sediment. These layers of sediment then turn into a rock through a process called *lithification.*

Sediments go through compaction and cementation during lithification. *Compaction* happens when the individual pieces of sediment in a layer are forced together because of the combined weight of all the sediment in layers above them. *Cementation* happens when the dissolved minerals between the pieces of sediment dry. These minerals then bind the other pieces of sediment together and harden—much like cement mix does after water is added to it. The sediments that are cemented together to create a sedimentary rock often come from all different kinds of rocks, and therefore have different physical properties.

LAB 10

Geologists, as a result, classify clastic sedimentary rocks based on the types of sediment found within them.

Sedimentary rocks, such as those shown in Figure L10.1, consist of layers of different types of sediments because different types of sediments fall through a fluid, such as water, at different rates. The rate at which a sediment falls through a fluid is called its *settling velocity*. The settling velocity of a sediment, like its texture, density, or color, is a unique physical property of that sediment. When a sediment has a settling velocity that is lower than the stream flow velocity of a river, that sediment will be carried downstream. When a sediment has a settling velocity that is higher than the stream flow velocity of a river, that sediment will sink to the bottom of the river and not move farther downstream. The stream velocity of a river at different locations, as a result, will determine where different types of sediments will accumulate and what types of sedimentary rocks will form at different locations.

It is important for geologists to understand how the different characteristics of a sediment affects its settling velocity because this physical property helps them explain how sediments move from one location to another and allows them to predict where different types of sediments will accumulate over time. Geologists can also learn more about environmental conditions of the past if they understand the factors that affect the deposition of sediments when they examine the nature and location of different types of sedimentary rock.

A sediment has many different physical properties that may or may not affect its settling velocity; these properties include particle size, shape, and density. Sediment particles can range in size from clay that is less than to 0.002 mm in diameter to large pebbles that can be well over 4 mm in diameter. Geologists often use a specific scale, such as the Wentworth scale (see Table L10.1), to classify or describe the particle size of a sediment. Shape is another physical property of a sediment, and geologists often classify or describe the shape of a sediment particle using the terms shown in Figure L10.2. Finally, the density of a sediment is defined as its mass per unit volume.

In this investigation, you will have an opportunity to figure out how these three physical properties affect the settling velocity of a sediment. You will then use this information to develop a conceptual model that explains how sediments will

TABLE L10.1

Modified Wentworth scale for classifying particles by size

Name	Particle size (mm)
Pebble	> 4
Granule	3.9–2.0
Very coarse sand	1.9–1.0
Coarse sand	0.9–0.5
Medium sand	0.49–0.25
Fine sand	0.24–0.125
Very fine sand	0.124–0.0625
Silt	0.0624–0.002
Clay	< 0.002

FIGURE L10.2

The various shapes of a sediment particle

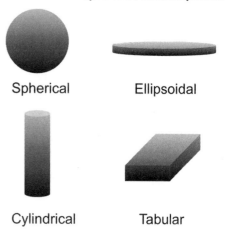

Spherical

Ellipsoidal

Cylindrical

Tabular

settle in water. A conceptual model like this is useful because it allows people to not only understand why sedimentary rocks consist of layers of different materials but also predict the order of the layers that will be found in different kinds of sedimentary rocks.

Your Task

You will be given several different sediments. You will then explore how each type of sediment settles in a column of water over time. Your goal is to use what you know about cause-and-effect relationships and structure and function to design and carry out an investigation that will enable you to develop a conceptual model that explains how particle size, density, and shape affects the settling rate of a sediment in water. Once you have created your conceptual model, you will return these sediments to your teacher. He or she will then give you one or two sediment mixtures. You will use the sediment mixtures to test, and if needed, revise your model. Your model, if valid or acceptable, should allow you to predict how the different types of sediments found in each mixture will accumulate at the bottom of a column of water over time.

The guiding question of this investigation is, *How can we explain the deposition of sediments in water?*

Materials

You may use any of the following materials during your investigation:

Sediments of different sizes

- Gravel
- Coarse sand
- Medium sand
- Fine sand

Sediments of different shapes

- 4 Pieces of modeling clay (each 2 g)

Sediments of different densities

- Glass beads (4 mm)
- Plastic beads (4 mm)
- Ball bearings (4 mm)

Sediment mixtures for testing the model

- Mixture A
- Mixture B

Consumable

- Water

Equipment

- Safety glasses or goggles (required)
- Chemical-resistant apron (required)
- Gloves (required)
- Clear plastic tube
- Rubber stopper
- Beaker (500 ml)
- Beaker (50 ml)
- Funnel
- Duct tape
- Meterstick
- Bucket
- Electronic or triple beam balance
- Stopwatches

Safety Precautions

Follow all normal lab safety rules. In addition, take the following safety precautions:

LAB 10

- Wear sanitized indirectly vented chemical-splash goggles and chemical-resistant aprons and gloves throughout the entire investigation (which includes setup and cleanup).
- Handle all glassware with care.
- Immediately wipe up any spilled water and pick up any spilled beads, ball bearings, or other materials that can cause a slip, trip, or fall hazard.
- Wash hands with soap and water when done collecting the data and after completing the lab.

Investigation Proposal Required?   ☐ Yes   ☐ No

Getting Started

You will need to design and carry out at least three different experiments to determine how the structure of a sediment affects the rate at which it will fall through a column of water. These experiments are necessary because you will need to answer three specific questions before you can develop an answer to the guiding question for this lab:

- How does particle size affect the time it takes a sediment to fall through a column of water?
- How does particle density affect the time it takes a sediment to fall through a column of water?
- How does particle shape affect the time it takes a sediment to fall through a column of water?

You can create a water column, such as the one shown in Figure L10.3, using a large plastic tube and a rubber stopper. Before you create your water column, it will be important for you to determine what type of data you need to collect, how you will collect the data, and how you will analyze the data for each experiment, because each experiment is slightly different.

To determine *what type of data you need to collect,* think about the following questions:

- What conditions need to be satisfied to establish a cause-and-effect relationship?
- How will you determine when a particular sediment type has settled?
- How will you determine how long it takes for a sediment to fall through a column of water?
- What information will you need to calculate the density of a particle?

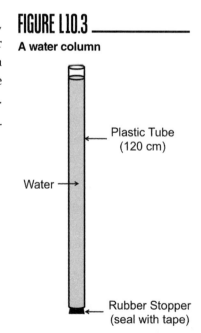

FIGURE L10.3

A water column

Plastic Tube (120 cm)

Water →

Rubber Stopper (seal with tape)

- When will you need to make these measurements or observations?

To determine *how you will collect the data,* think about the following questions:

- What will serve as your independent variable and dependent variable for each experiment?
- How will you vary the independent variable during each experiment?
- What will you do to hold the other variables constant during each experiment?
- What types of comparisons will you need to make?
- What measurement scale or scales should you use to collect data?
- What will you do to reduce error in your measurements?
- How will you keep track of and organize the data you collect?

To determine *how you will analyze the data,* think about the following questions:

- What calculations will you need to make?
- How could you use mathematics to describe a relationship between variables?
- How could you use mathematics to document a difference between groups or conditions?
- What types of patterns might you look for as you analyze your data?
- What type of table or graph could you create to help make sense of your data?

Once you have finished collecting data, your group can develop a conceptual model that explains the deposition of sediments in water. For your conceptual model to be complete, it must be able to explain how the structure of different sediments relates to how different types of sediments will move through water. It must also include information about the stream flow velocity of a body of water such as a river or lake. The stream flow velocity is how fast the water is moving in a specific direction. The stream flow velocity of water in a water column is zero, but in a river it can reach velocities of 25 km/h. Finally, and perhaps most important, you should be able to use your model to predict the order in which different types of sediments will settle at the bottom of a column of water. This type of conceptual model is useful because it enables people to understand where and when different types of sediments will accumulate over time.

The last step in your investigation will be to generate the evidence that you need to convince others that the conceptual model you developed based on your experiments is valid or acceptable. To accomplish this goal, you can use your model to predict how the different sediments in a mixture will settle in a column of water after a set amount of time. If you are able to use your conceptual model to make accurate predictions about how the different sediments in a mixture will move through the water relative to each other, then you should be able to convince others that your model is valid or acceptable.

LAB 10

Connections to the Nature of Scientific Knowledge and Scientific Inquiry

As you work through your investigation, be sure to think about

- the use of models as as tools for reasoning about natural phenomena, and
- the nature and role of experiments in science.

Initial Argument

Once your group has finished collecting and analyzing your data, your group will need to develop an initial argument. Your initial argument needs to include a claim, evidence to support your claim, and a justification of the evidence. The *claim* is your group's answer to the guiding question. The *evidence* is an analysis and interpretation of your data. Finally, the *justification* of the evidence is why your group thinks the evidence matters. The justification of the evidence is important because scientists can use different kinds of evidence to support their claims. Your group will create your initial argument on a whiteboard. Your whiteboard should include all the information shown in Figure L10.4.

FIGURE L10.4 _____

Argument presentation on a whiteboard

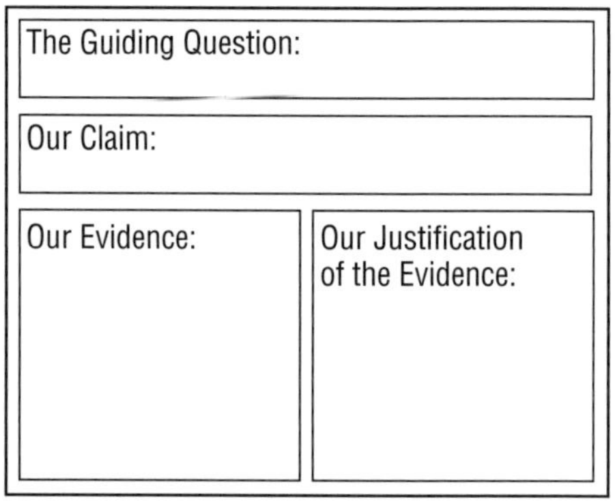

The Guiding Question:	
Our Claim:	
Our Evidence:	Our Justification of the Evidence:

Argumentation Session

The argumentation session allows all of the groups to share their arguments. One or two members of each group will stay at the lab station to share that group's argument, while the other members of the group go to the other lab stations to listen to and critique the other arguments. This is similar to what scientists do when they propose, support, evaluate, and refine new ideas during a poster session at a conference. If you are presenting your group's argument, your goal is to share your ideas and answer questions. You should also keep a record of the critiques and suggestions made by your classmates so you can use this feedback to make your initial argument stronger. You can keep track of specific critiques and suggestions for improvement that your classmates mention in the space below.

Critiques of our initial argument and suggestions for improvement:

If you are critiquing your classmates' arguments, your goal is to look for mistakes in their arguments and offer suggestions for improvement so these mistakes can be fixed. You should look for ways to make your initial argument stronger by looking for things that the other groups did well. You can keep track of interesting ideas that you see and hear during the argumentation in the space below. You can also use this space to keep track of any questions that you will need to discuss with your team.

Interesting ideas from other groups or questions to take back to my group:

Once the argumentation session is complete, you will have a chance to meet with your group and revise your initial argument. Your group might need to gather more data or design a way to test one or more alternative claims as part of this process. Remember, your goal at this stage of the investigation is to develop the best argument possible.

Report

Once you have completed your research, you will need to prepare an *investigation report* that consists of three sections. Each section should provide an answer for the following questions:

1. What question were you trying to answer and why?

2. What did you do to answer your question and why?

3. What is your argument?

Your report should answer these questions in two pages or less. You should write your report using a word processing application (such as Word, Pages, or Google Docs), if possible, to make it easier for you to edit and revise it later. You should embed any diagrams, figures, or tables into the document. Be sure to write in a persuasive style; you are trying to convince others that your claim is acceptable or valid.

LAB 10

Checkout Questions

Lab 10. Deposition of Sediments: How Can We Explain the Deposition of Sediments in Water?

1. Use numbers to rank the following sediments from the greatest to smallest likely settling velocity. Assume that the sediments all have the same density. If you think any two sediments will have the same settling velocity, give them the same number.

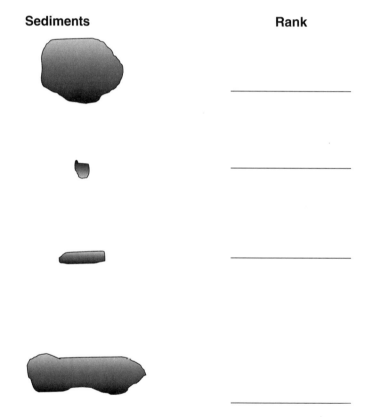

Sediments   **Rank**

_____

_____

_____

_____

Explain your answer. Why do you think the order that you chose is correct?

2. Jalen adds a sample of soil to a jar. The soil is made up of fine sand, coarse sand, gravel, and clay. He then fills the remaining space in the jar with water, leaving a little room for air at the top. After putting on the lid, he shakes the jar until all the soil particles are mixed with the water. He leaves the jar on a table overnight. The next morning, he sees four layers of sediment in the jar. Which picture shows how you think the different types of soil particles will settle in the jar: A, B, or C?

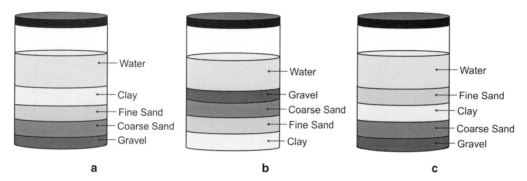

Explain your answer. Why do you think the order that you chose is correct?

3. Scientists often develop models to help them understand complex phenomena.

   a. I agree with this statement.

   b. I disagree with this statement.

   Explain your answer, using an example from your investigation about the deposition of sediments.

4. Scientists use experiments to prove ideas right or wrong.

    a. I agree with this statement.

    a. I disagree with this statement.

Explain your answer, using an example from your investigation about the deposition of sediments.

5. Scientists are often interested in identifying cause-and-effect relationships. Explain what a cause-and-effect relationship is and why these relationships are important, using an example from your investigation about the deposition of sediments.

6. In nature, the way something is structured often determines its function or places limits on what it can or cannot do. Explain why it is important to keep in mind the relationship between structure and function when attempting to collect or analyze data, using an example from your investigation about the deposition of sediments.

Lab Handout

Lab 11. Soil Texture and Soil Water Permeability: How Does Soil Texture Affect Soil Water Permeability?

FIGURE L11.1 _____

Soil particle types and sizes

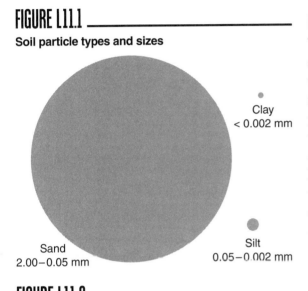

Clay
< 0.002 mm

Silt
0.05–0.002 mm

Sand
2.00–0.05 mm

FIGURE L11.2 _____

Soil types by composition

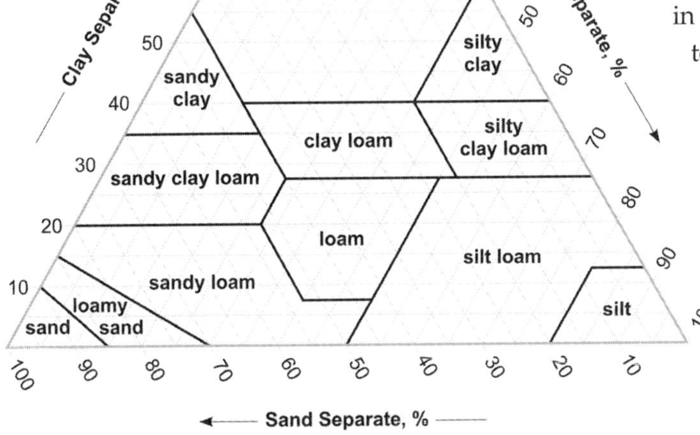

Introduction

When it rains, water lands on the ground and flows into streams, rivers, gutters, and sewers. It can also soak into the ground. The ground is made up rock and soil. Soil is composed of many small particles. In between these particles are pores. Water can flow into and through these pores. *Soil water permeability* is defined as the rate at which water can flow through the pores of a soil. Rainwater tends to seep into soils that have high soil water permeability and tends to pool on top of or flow over soils with low soil water permeability. There are several different physical properties of soil that can affect the rate at which water flows through it.

One of the physical properties of soil that affects soil water permeability is *soil texture.* Soil is composed of small particles that vary in size, shape, and chemical composition. There are three main types of soil particles (see Figure L11.1): (a) sand particles are between 0.05 mm and 2.0 mm in diameter, (b). silt particles are between 0.002 mm and 0.05 mm in diameter, and (c). clay particles are smaller than 0.002 mm. *Soil texture* describes the relative proportion of sand, silt, and clay particles by weight in soil sample. Earth scientists use a soil texture triangle, such as the one shown in Figure L11.2, to classify soils according to 12 textural categories, based on the proportions of each particle type in the soil sample. For example, a soil that is composed of 30% clay, 10% silt, and 60% sand is classified as a *sandy clay loam;* whereas a soil that consists of 20% clay, 40% silt and 40% sand *is classified as a* loam.

Geotechnical engineers are often concerned about the water permeability of soil because they

LAB 11

are responsible for building structures on or in the ground, and the rate at which water flows through soil can have an adverse effect on these structures. Geotechnical engineers must therefore understand how the physical properties of soil affect the rate at which water flows through it. In this investigation, you will have an opportunity to examine how soil texture, which is an important physical property of soil, affects soil water permeability.

Your Task

Use what you know about soil composition; scale, proportion, and quantity; and the importance of tracking how matter moves into, through, and out of a system to plan and carry out an investigation to determine the relationship between soil texture and soil water permeability.

The guiding question of this investigation is, *How does soil texture affect soil water permeability?*

Materials

You may use any of the following materials during your investigation:

Consumables
- Soil sample A
- Soil sample B
- Soil sample C
- Soil sample D
- 12 Coffee filters
- Paper towels

Equipment
- Safety glasses or goggles (required)
- Chemical-resistant apron (required)
- Gloves (required)
- Measuring spoon (15 ml)
- 1 Graduated cylinder (100 ml)
- 1 Graduated cylinder (50 ml)
- 1–4 Funnels
- 1–4 Beakers (each 250 ml)
- 1–4 Support stands
- 1–4 Ring clamps

- 1 Stopwatch
- 1 Wax marking pencil
- Electronic or triple beam balance
- Flowchart for soil texture by feel

Safety Precautions

Follow all normal lab safety rules. In addition, take the following safety precautions:

- Wear sanitized indirectly vented chemical-splash goggles and chemical-resistant, nonlatex aprons and gloves through the entire investigation (which includes setup and cleanup).

- Handle all glassware with care.

- Report and clean up any spills immediately, and avoid walking in areas where water has been spilled.

- Wash hands with soap and water when done collecting the data and after completing the lab.

Investigation Proposal Required?   ☐ Yes   ☐ No

Getting Started

You can determine the rate at which water flows through a soil sample using the equipment illustrated in Figure L11.3. Once you have your equipment set up, you can place a sample of soil inside the coffee filter and then add water to the soil. The water will flow through the soil and the coffee filter and then land in the beaker. Your teacher will let you know where you can get samples of different types of soil.

Before you begin to design your experiment using this equipment, think about what type of data you need to collect, how you will collect the data, and how you will analyze the data.

FIGURE L11.3

Equipment needed to measure the rate at which water flows through a sample of soil

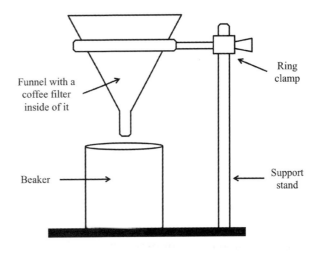

To determine *what type of data you need to collect,* think about the following questions:

- How can you track water flowing through soil?
- What information about a soil sample do you need to determine its texture?
- What information will you need to calculate a rate?
- What type of measurements or observations will you need to record during each experiment?
- When will you need to make these measurements or observations?

To determine *how you will collect the data,* think about the following questions:

- What will serve as your independent variable and dependent variable?
- How will you vary the independent variable during each experiment?
- What will you do to hold the other variables constant during each experiment?
- What types of comparisons will you need to make?
- What measurement scale or scales should you use to collect data?
- What will you do to reduce error in your measurements?
- How will you keep track of and organize the data you collect?

To determine *how you will analyze the data,* think about the following questions:

- What calculations will you need to make?
- How could you use mathematics to describe a relationship between variables?
- How could you use mathematics to document a difference between groups or conditions?
- What types of patterns might you look for as you analyze your data?
- What type of table or graph could you create to help make sense of your data?

Connections to the Nature of Scientific Knowledge and Scientific Inquiry

As you work through your investigation, be sure to think about

- the difference between observations and inferences in science, and
- how the culture of science, societal needs, and current events influence the work of scientists.

Initial Argument

Once your group has finished collecting and analyzing your data, your group will need to develop an initial argument. Your initial argument needs to include a claim, evidence to support your claim, and a justification of the evidence. The *claim* is your group's answer to the guiding question. The *evidence* is an analysis and interpretation of your data. Finally, the *justification* of the evidence is why your group thinks the evidence matters. The justification of the evidence is important because scientists can use different kinds of evidence to support their claims. Your group will create your initial argument on a whiteboard. Your whiteboard should include all the information shown in Figure L11.4.

FIGURE L11.4 _____

Argument presentation on a whiteboard

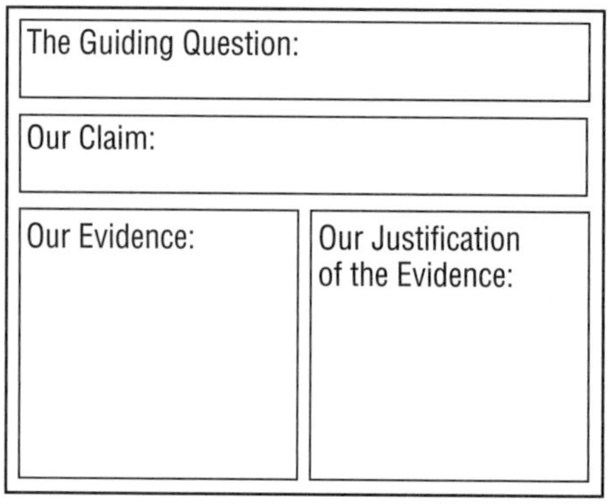

The Guiding Question:	
Our Claim:	
Our Evidence:	Our Justification of the Evidence:

Argumentation Session

The argumentation session allows all of the groups to share their arguments. One or two members of each group will stay at the lab station to share that group's argument, while the other members of the group go to the other lab stations to listen to and critique the other arguments. This is similar to what scientists do when they propose, support, evaluate, and refine new ideas during a poster session at a conference. If you are presenting your group's argument, your goal is to share your ideas and answer questions. You should also keep a record of the critiques and suggestions made by your classmates so you can use this feedback to make your initial argument stronger. You can keep track of specific critiques and suggestions for improvement that your classmates mention in the space provided.

Critiques of our initial argument and suggestions for improvement:

If you are critiquing your classmates' arguments, your goal is to look for mistakes in their arguments and offer suggestions for improvement so these mistakes can be fixed. You should look for ways to make your initial argument stronger by looking for things that the other groups did well. You can keep track of interesting ideas that you see and hear during the argumentation in the space below. You can also use this space to keep track of any questions that you will need to discuss with your team.

Interesting ideas from other groups or questions to take back to my group:

LAB 11

Once the argumentation session is complete, you will have a chance to meet with your group and revise your initial argument. Your group might need to gather more data or design a way to test one or more alternative claims as part of this process. Remember, your goal at this stage of the investigation is to develop the best argument possible.

Report

Once you have completed your research, you will need to prepare an *investigation report* that consists of three sections. Each section should provide an answer for the following questions:

1. What question were you trying to answer and why?

2. What did you do to answer your question and why?

3. What is your argument?

Your report should answer these questions in two pages or less. You should write your report using a word processing application (such as Word, Pages, or Google Docs), if possible, to make it easier for you to edit and revise it later. You should embed any diagrams, figures, or tables into the document. Be sure to write in a persuasive style; you are trying to convince others that your claim is acceptable or valid.

Checkout Questions

Lab 11. Soil Texture and Soil Water Permeability: How Does Soil Texture Affect Soil Water Permeability?

1. Peyton has collected some soil samples. The figure below shows a microscopic view of his three different soils. Assume each sample is viewed under the same magnification.

O = .002 mm

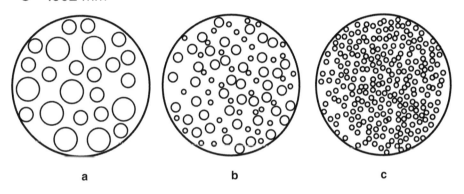

a   b   c

a. Which soil is made mostly of clay?

b. How do you know?

c. Which soil has the highest rate of soil water permeability?

d. How do you know?

2. Jackson has a sample of soil. He sets up a ring stand and a funnel with a filter and places a beaker below. He places 100 ml of water into the soil and waits for the water to flow through it. After two minutes, 40 ml of water is in the beaker. What is the rate of soil water permeability of Jackson's soil sample?

3. Scientists use proportional relationships to classify matter.

   a. I agree with this statement.

   b. I disagree with this statement.

   Explain your answer, using an example from your investigation about soil water permeability.

4. Adrianna identified one of her soil samples as sandy clay loam using the flowchart for soil texture by feel. She tells her group that she has made an inference about the sample's texture.

   a. I agree with this statement.

   b. I disagree with this statement.

   Explain your answer, using an example from your investigation about soil water permeability.

5. Scientists are influenced by many different factors when doing their work. The culture of science represents a shared set of values, norms, and commitments that shape what counts as knowing something in science, how to represent and communicate information, and how to interact with other scientists. How did your group decide that your claim was valid scientific knowledge? Give an example from your investigation about soil porosity.

6. Scientists often track how matter moves into, out of, and within systems during an investigation. Explain why it useful to track the movement of matter into, out of, and within a system using an example from your investigation about soil water permeability.

Lab Handout

Lab 12. Cycling of Water on Earth: Why Do the Temperature and the Surface Area to Volume Ratio of a Sample of Water Affect Its Rate of Evaporation?

Introduction

Water can be found as a liquid, solid, or gas on Earth. Lakes, rivers, and oceans contain liquid water. For example, Lake Tahoe (shown in Figure L12.1a), which straddles the border between California and Nevada, contains 36.15 cubic miles of liquid water. Polar ice caps and glaciers contain solid water. For example, the Perito Moreno glacier (see Figure L12.1b), located in western Patagonia, Argentina, is a huge ice formation that is 30 km in length, 5 km wide, and has an average height of 74 m. The atmosphere contains gaseous water vapor. Unlike the other two forms of water, gaseous water vapor is invisible. We can see water vapor only when it condenses to form visible clouds of water droplets such as in Figure L12.1c. Water vapor is responsible for humidity.

FIGURE L12.1 _____

Water can be found in all three states on Earth. (a) Lake Tahoe contains liquid water. (b) The Perito Moreno glacier contains solid water. (c) Gaseous water is invisible unless it condenses to form clouds of water droplets, as seen in the Owakudani volcanic valley in Japan.

a   b   c

All water is made up of molecules that are composed of two hydrogen atoms and one oxygen atom. This type of molecule is called a water molecule. All water molecules have the same mass. Water molecules are also constantly in motion. Because water molecules have mass and are constantly in motion, they have *kinetic energy*. Temperature is a way we can measure the average kinetic energy of the molecules in any sample of water. A high temperature means that the molecules in the sample have high average kinetic energy and are moving quickly, while a low temperature means the molecules have low kinetic energy and are moving slowly. The three states of water are determined by temperature. At low temperatures (less than 0°C), water is a solid (ice); at room temperature, water is a liquid; and at high temperatures (more than 100°C), water is a gas.

FIGURE L12.2

The water cycle explains how water molecules move into, out of, and within Earth's systems

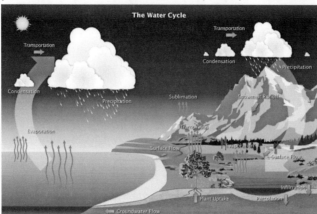

The water cycle, shown in Figure L12.2, is a model that scientists use to explain how water molecules move into, out of, and within Earth's systems. This process is driven by energy from the Sun and the force of gravity. When energy from the Sun heats liquid water, some of it transforms into gaseous water vapor and enters the atmosphere. This process is called evaporation. The water vapor rises into the air, where cooler temperatures cause it to condense into tiny liquid water droplets. A huge concentration of these droplets becomes visible to us as a cloud. Air currents move clouds around the globe. The water droplets in clouds collide, grow, and eventually fall to the ground as precipitation. Some precipitation falls as snow and can accumulate as ice caps and glaciers, which can store frozen water for thousands of years. Snowpacks in warmer climates often thaw and melt in the spring, and the melted water flows overland as snowmelt. Most precipitation, however, falls back into the oceans or onto land. On land, the water will flow over the ground as surface runoff because of the force of gravity.

A portion of the runoff will enter rivers in valleys in the landscape and move toward an ocean. Some of the runoff accumulates and forms freshwater lakes. Some of the water will also soak into the ground. This water is called groundwater. Some of the groundwater will move deep into the ground and create aquifers, which are underground stores of freshwater, and some will stay close to the surface. The groundwater that remains close to the surface will seep back into lakes or rivers as groundwater discharge or will create freshwater springs. Over time, though, all of this water keeps moving, and it eventually reaches an ocean or returns to the atmosphere through the process of evaporation.

Water can evaporate and enter the atmosphere from sources as vast as the ocean and as small as your pet's water dish. At any place where the surface of the water meets with the air, water molecules are able to leave the liquid water and enter the atmosphere. It might seem like evaporation makes liquid water disappear, but recall that the law of conservation of matter states that matter can never be created nor destroyed, but it can change form. When evaporation happens, the molecules are simply changing from a liquid phase, which is visible to us, to a gaseous phase, which we cannot see. Keeping this in mind, we can tell how much evaporation has happened by measuring changes in the mass or volume of the liquid water.

You may have noticed that bodies of water evaporate at different rates. For example, a rain puddle on the street can evaporate in a few hours, but water in a glass may take days or weeks to evaporate. There are many factors that may affect the rate that water evaporates:

- The amount of energy that water absorbs from the Sun
- The temperature of the water
- The *surface area to volume ratio* of the water; or the amount of the water's surface that is exposed to the air compared with its total volume

In this investigation, you will have an opportunity to determine how water temperature and the surface area to volume ratio of a sample of water contribute to the rate that water evaporates. Once you understand how these two factors affect how quickly a sample of water will evaporate, you will then develop a conceptual model that you can use to explain your observations and predict how quickly water will evaporate under different conditions.

Your Task

Use what you know about the properties of water, rates of change, and the importance of tracking the movement of matter into, out of, and within systems during an investigation to plan and carry out an experiment to determine how changes in the temperature and the surface area to volume ratio of a sample of water affect how quickly it will evaporate. Then develop a conceptual model that can be used to explain *why* these factors affect the rate that water evaporates. Once you have developed your conceptual model, you will need to test it using different water samples to determine if it allows you to predict how much liquid water will be lost due to evaporation under different conditions.

The guiding question of this investigation is, **Why do the temperature and the surface area to volume ratio of a sample of water affect its rate of evaporation?**

LAB 12

Materials

You may use any of the following materials during your investigation:

Consumables
- Water
- Ice

Equipment
- Safety glasses or goggles (required)
- Chemical-resistant apron (required)
- Pyrex containers of different shapes and sizes
- Electronic or triple beam balance
- Graduated cylinder (250 ml)
- Graduated cylinder (100 ml)
- Beaker (500 ml)
- Beaker (250 ml)
- Beaker (150 ml)
- Thermometer (nonmercury)
- 2 Heat lamps
- Hot plate
- Ruler

Safety Precautions

Follow all normal lab safety rules. In addition, take the following safety precautions:

- Wear sanitized indirectly vented chemical-splash goggles and chemical-resistant, nonlatex aprons throughout the entire investigation (which includes setup and cleanup).
- Report and clean up spills immediately, and avoid walking in areas where water has been spilled
- Use caution when working with heat lamps and hot plates; they can get hot enough to burn skin.
- Never spray water on hot heat lamps—glass will shatter, producing dangerous glass projectiles.
- Use only GFCI-protected electrical receptacles for heat lamps and hot plates to prevent or reduce potential shock hazard.
- Handle all glassware with care.
- Handle glass thermometers with care. They are fragile and can break, causing a sharp hazard that can cut or puncture skin.
- Wash hands with soap and water when done collecting the data and after completing the lab.

Investigation Proposal Required?   ☐ Yes   ☐ No

Getting Started

The first step in developing your model is to plan and carry out at least two experiments. Figure L12.3 shows how you can use a heat lamp to warm your different samples of water. The heat lamp will serve as the source of energy for each experiment. Your teacher may also allow you to set your water samples outside in direct sunlight depending on the time of year. Figure L12.3 also shows how you can use containers of different sizes and shapes to manipulate the surface area to volume ratio of a sample of water. You can use a hot plate to heat your water samples to different temperatures or to maintain the temperature of a water sample.

Before you begin to design your two experiments using this equipment, be sure to think about what type of data you need to collect, how you will collect the data, and how you will analyze the data. To determine *what type of data you need to collect,* think about the following questions:

- How will you track the flow of energy into each water sample?
- How will you track the amount of water loss from a sample?
- How will you measure the rate of water evaporation (change over time)?

FIGURE L12.3 _____

How to use a heat lamp to warm different samples of water

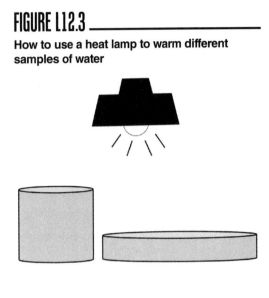

To determine *how you will collect the data,* think about the following questions:

- What will be the independent and dependent variables for each experiment?
- What conditions will you need to set up for each experiment?
- How will you make sure you are only testing one variable at a time?
- How often will you need to take measurements during each experiment?
- What measurement scale or scales should you use to collect data?
- How will you make sure that your data are of high quality (i.e., how will you reduce error)?
- How will you keep track of and organize the data you collect?

To determine *how you will analyze the data,* think about the following questions:

- What type of calculations will you need to make?
- How will you determine if rates of change are the same or different?
- How could you use mathematics to document a difference between conditions?
- What type of table or graph could you create to help make sense of your data?

Once you have carried out your two experiments, you will need to develop a conceptual model. Your model needs to be able to explain why temperature and the surface area to volume ratio of a sample of water affect the amount of water lost due to evaporation in the way that these two factors do. The model also needs to account for the kinetic energy of water molecules and the conservation of matter.

The last step in this investigation is to test your model. To accomplish this goal, you can set up a third experiment to determine if your model leads to accurate predictions about the amount of water that will be lost from different containers of water under specific conditions (e.g., cold water, high surface area to volume ratio). If you are able to use your model to make accurate predictions about the way water evaporates under different conditions, then you will be able to generate the evidence you need to convince others that the conceptual model you developed is valid or acceptable.

Connections to the Nature of Scientific Knowledge and Scientific Inquiry

As you work through your investigation, be sure to think about

- the difference between observations and inferences in science, and
- the nature and role of experiments in science.

Initial Argument

Once your group has finished collecting and analyzing your data, your group will need to develop an initial argument. Your initial argument needs to include a claim, evidence to support your claim, and a justification of the evidence. The *claim* is your group's answer to the guiding question. The *evidence* is an analysis and interpretation of your data. Finally, the *justification* of the evidence is why your group thinks the evidence matters. The justification of the evidence is important because scientists can use different kinds of evidence to support their claims. Your group will create your initial argument on a whiteboard. Your whiteboard should include all the information shown in Figure L12.4.

FIGURE L12.4

Argument presentation on a whiteboard

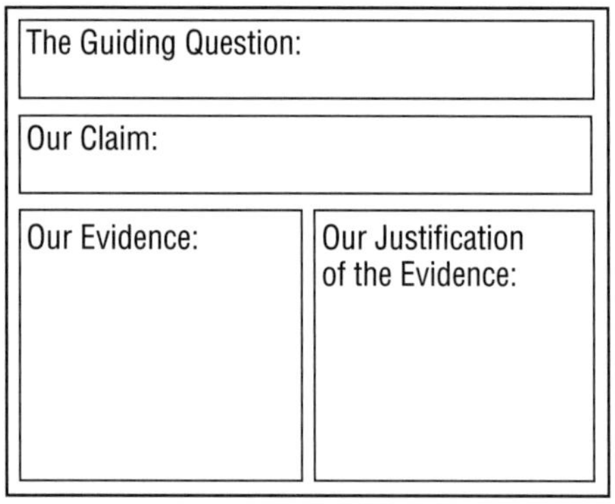

The Guiding Question:	
Our Claim:	
Our Evidence:	Our Justification of the Evidence:

Argumentation Session

The argumentation session allows all of the groups to share their arguments. One or two members of each group will stay at the lab station to share that group's argument, while the other members of the group go to the other lab stations to listen to and critique the other arguments. This is similar to what scientists do when they propose, support, evaluate, and refine new ideas during a poster session at a conference. If you are presenting your group's argument, your goal is to share your ideas and answer questions. You should also keep a record of the critiques and suggestions made by your classmates so you can use this feedback to make your initial argument stronger. You can keep track of specific critiques and suggestions for improvement that your classmates mention in the space below.

Critiques of our initial argument and suggestions for improvement:

If you are critiquing your classmates' arguments, your goal is to look for mistakes in their arguments and offer suggestions for improvement so these mistakes can be fixed. You should look for ways to make your initial argument stronger by looking for things that the other groups did well. You can keep track of interesting ideas that you see and hear during the argumentation in the space below. You can also use this space to keep track of any questions that you will need to discuss with your team.

Interesting ideas from other groups or questions to take back to my group:

Once the argumentation session is complete, you will have a chance to meet with your group and revise your initial argument. Your group might need to gather more data or design a way to test one or more alternative claims as part of this process. Remember, your goal at this stage of the investigation is to develop the best argument possible.

Report

Once you have completed your research, you will need to prepare an investigation report that consists of three sections. Each section should provide an answer for the following questions:

1. What question were you trying to answer and why?

2. What did you do to answer your question and why?

3. What is your argument?

Your report should answer these questions in two pages or less. You should write your report using a word processing application (such as Word, Pages, or Google Docs), if possible, to make it easier for you to edit and revise it later. You should embed any diagrams, figures, or tables into the document. Be sure to write in a persuasive style; you are trying to convince others that your claim is acceptable or valid.

Checkout Questions

Lab 12. Cycling of Water on Earth: Why Do the Temperature and the Surface Area to Volume Ratio of a Sample of Water Affect Its Rate of Evaporation?

1. How does the temperature of a water sample affect the rate of its evaporation?

2. How does the surface area to volume ratio of a water sample affect the rate of its evaporation?

3. A scientist has a collection of water samples. Each sample has the same volume of water, but the samples vary in their surface areas and temperatures. She is trying to compare the amount of evaporation in each sample.

Sample	Sample surface area (cm²)	Volume (cm³)	Sample temperature (°C)
A	6.25	20.0	25
B	10.00	20.0	65
C	12.56	20.0	25
D	3.14	20.0	65

a. After 45 minutes, which sample will have evaporated the most?

b. How do you know?

c. After 45 minutes, which sample will have evaporated the least?

d. How do you know?

4. This investigation was an experiment.

    a. I agree with this statement.

    b. I disagree with this statement.

Explain your answer, using an example from your investigation about the cycling of water on Earth.

5. "Sample A lost 5 grams of water due to evaporation" is an example of an observation.

    a. I agree with this statement.

    b. I disagree with this statement.

Explain your answer, using an example from your investigation about the cycling of water on Earth.

6. Scientists often track how matter moves into, out of, and within systems during an investigation. Explain why it is useful to do this, using an example from your investigation about the cycling of water on Earth.

7. Scientists often try to understand what controls the rate of change of a system. Explain what a rate of change in a system is and why it is useful to understand the factors that control a rate of change in a system, using an example from your investigation about the cycling of water on Earth.

Application Labs

Lab Handout

Lab 13. Characteristics of Minerals: What Are the Identities of the Unknown Minerals?

Introduction

Rocks are made up of different types of minerals or other pieces of rock, which are made of minerals. Granite (Figure L13.1) and marble (Figure L13.2) are examples of different kinds of rock. Earth scientists group rocks into one of three categories:

- *Sedimentary* rocks are formed at the Earth's surface by the accumulation and cementation of fragments of sediments. Sandstone is an example of a sedimentary rock.

- *Igneous* rocks form through the cooling and solidification of magma or lava. Granite is an example of an igneous rock.

- *Metamorphic* rocks are produced when existing rocks are subjected to extreme temperature and pressure. Marble is an example of a metamorphic rock.

FIGURE L13.1 _____

A granite outcrop at Logan Rock, Cornwall, England

FIGURE L13.2 _____

Marble at a quarry in Carrara, Italy

Earth scientists use the mineral composition of a rock to classify it. Granite, for example, is made up minerals such as quartz, feldspar, and biotite; marble is made up of minerals called dolomite and calcite. Earth scientists must be able to determine the various types of minerals that are in a rock in order to identify it. Every mineral has a unique chemical composition. Dolomite (see Figure L13.3), for example, has the chemical composition of $CaMg(CO_3)_2$, whereas quartz (see Figure L13.4) has a chemical composition of SiO_2. The unique chemical composition of a mineral gives it a specific combination of chemical and physical properties. Earth scientists use these chemical and physical properties to identify a mineral.

FIGURE L13.3 _____

Dolomite

FIGURE L13.4 _____

Quartz

Chemical properties (see Figure L13.5) describe how a mineral interacts with other types of matter. Dolomite, for example, reacts with hydrochloric acid but quartz does not. *Physical properties* are descriptive characteristics of a mineral. Examples of physical properties include color, density, streak (whether the mineral streaks on a streak plate and the color of the powder), hardness (whether the mineral can scratch something with a known hardness, like glass or a nail), smell, how the mineral breaks (*cleavage* is when a mineral breaks evenly along a flat surface; *fracture* is when a mineral breaks apart roughly), and luster (whether the material appears metallic or nonmetallic). Some minerals will even attract magnets.

It is often challenging to determine the identity of an unknown mineral based on its chemical and physical properties. For example, if an Earth scientist has only a small amount of a mineral, he or she may not be able to conduct all the different types of tests that are needed because some tests may change the characteristics of the mineral during the process (such as when dolomite is mixed with an acid). It is also difficult to determine

FIGURE L13.5 _____

How Earth scientists distinguish between different minerals

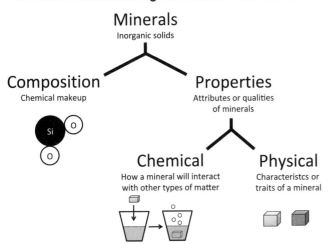

many of the physical properties of the sample, such as its density or its luster, when there is only a small amount of the substance, because taking measurements is harder. To complicate matters further, an unknown mineral may have an irregular shape, which can make it difficult to accurately measure its volume. Without knowing the mass and the volume of a substance, it is impossible to calculate its density. In this investigation, you will have an opportunity to learn about some of the challenges Earth scientists face when they need to identify an unknown mineral based on its chemical and physical properties and why it is important to make accurate measurements inside the laboratory.

Your Task

You will be given a set of known minerals. You will then document, measure, or calculate at least three different chemical or physical properties for each mineral. When you are done, you will return the known minerals to your teacher, who will then give you a set of unknown minerals. The set of unknowns will include samples of minerals that you tested. Your goal is to use what you know about the physical and chemical properties of matter, proportional relationships, and patterns to design and carry out an investigation that will enable you to collect the data you need to determine the identity of the unknown minerals.

The guiding question of this investigation is, *What are the identities of the unknown minerals?*

Materials

You may use any of the following materials during your investigation:

Consumables
- Water (in squirt bottles)
- Vinegar
- Set of known minerals
- Set of unknown minerals

Equipment
- Safety glasses or goggles (required)
- Chemical-resistant apron (required)
- Gloves (required)
- Electronic or triple beam balance
- Magnifying glass
- Magnet
- Beaker (250 ml)
- Beaker (400 ml)

- Graduated cylinder (100 ml)
- Overflow can
- Pipette
- Ruler
- Streak plate
- Small piece of glass
- Penny
- Nail

Safety Precautions

Follow all normal lab safety rules. In addition, take the following safety precautions:

- Follow safety precautions noted on safety data sheets for hazardous chemicals.
- Wear sanitized indirectly vented chemical-splash goggles and chemical-resistant, nonlatex aprons and gloves throughout the entire investigation (which includes setup and cleanup).

- Report and clean up any spills on the floor immediately to avoid a slip or fall hazard.
- Handle all glassware with care.
- When using the glass or streak plate, do not pick the plate up with your hands. Place it flat on the table. Otherwise, it may cut your hand.
- Handle nails with care. Nail ends are sharp and can cut or puncture skin.
- Do not place any materials in or around your mouth.
- Handle all glassware with care.
- Wash hands with soap and water when done collecting the data and after completing the lab.

Investigation Proposal Required?   ☐ Yes   ☐ No

Getting Started

To answer the guiding question, you will need to make several systematic observations of the known and unknown minerals. To accomplish this task, you must determine what type of data you need to collect, how you will collect it, and how you will analyze it.

To determine *what type of data you need to collect*, think about the following questions:

- Which three properties will you focus on as you make your systematic observations? The properties you choose to focus on can be chemical ones (reactions with other substances) or physical ones (e.g., color, density, hardness, streak).
- What information do you need to determine or calculate each of the chemical or physical properties?
- How will you determine if the physical properties of the various objects are the same or different?

To determine *how you will collect the data*, think about the following questions:

- What equipment will you need to collect the data you need?
- How will you make sure that your data are of high quality (i.e., how will you reduce error)?
- How will you keep track of and organize the data you collect?

To determine *how you will analyze the data*, think about the following questions:

- How might the unique chemical composition of a mineral (structure) be related to its unique chemical and physical properties (function)?
- What types of patterns might you look for as you analyze your data?
- What type of calculations will you need to make?

- What type of table or graph could you create to help make sense of your data?

Connections to the Nature of Scientific Knowledge and Scientific Inquiry

As you work through your investigation, be sure to think about

- the difference between data and evidence in science, and
- how scientists use different types of methods to answer different types of questions.

Initial Argument

Once your group has finished collecting and analyzing your data, your group will need to develop an initial argument. Your initial argument needs to include a claim, evidence to support your claim, and a justification of the evidence. The *claim* is your group's answer to the guiding question. The *evidence* is an analysis and interpretation of your data. Finally, the *justification* of the evidence is why your group thinks the evidence matters. The justification of the evidence is important because scientists can use different kinds of evidence to support their claims. Your group will create your initial argument on a whiteboard. Your whiteboard should include all the information shown in Figure L13.6.

FIGURE L13.6

Argument presentation on a whiteboard

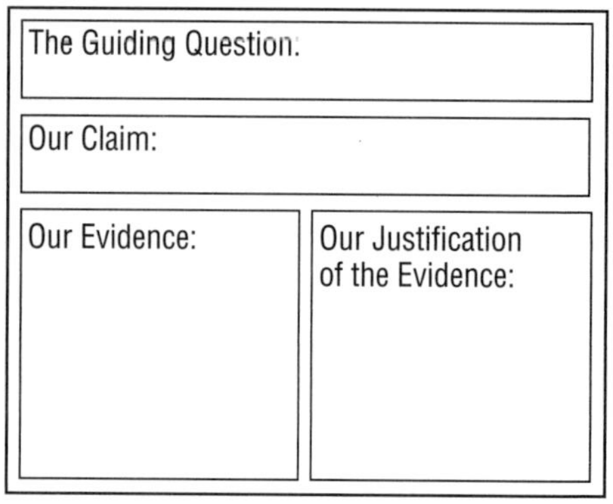

The Guiding Question:	
Our Claim:	
Our Evidence:	Our Justification of the Evidence:

Argumentation Session

The argumentation session allows all of the groups to share their arguments. One or two members of each group will stay at the lab station to share that group's argument, while the other members of the group go to the other lab stations to listen to and critique the other arguments. This is similar to what scientists do when they propose, support, evaluate, and refine new ideas during a poster session at a conference. If you are presenting your group's argument, your goal is to share your ideas and answer questions. You should also keep a record of the critiques and suggestions made by your classmates so you can use this feedback to make your initial argument stronger. You can keep track of specific critiques and suggestions for improvement that your classmates mention in the space below.

Critiques of our initial argument and suggestions for improvement:

If you are critiquing your classmates' arguments, your goal is to look for mistakes in their arguments and offer suggestions for improvement so these mistakes can be fixed. You should look for ways to make your initial argument stronger by looking for things that the other groups did well. You can keep track of interesting ideas that you see and hear during the argumentation in the space below. You can also use this space to keep track of any questions that you will need to discuss with your team.

Interesting ideas from other groups or questions to take back to my group:

LAB 13

Once the argumentation session is complete, you will have a chance to meet with your group and revise your initial argument. Your group might need to gather more data or design a way to test one or more alternative claims as part of this process. Remember, your goal at this stage of the investigation is to develop the best argument possible.

Report

Once you have completed your research, you will need to prepare an *investigation report* that consists of three sections. Each section should provide an answer for the following questions:

1. What question were you trying to answer and why?

2. What did you do to answer your question and why?

3. What is your argument?

Your report should answer these questions in two pages or less. You should write your report using a word processing application (such as Word, Pages, or Google Docs), if possible, to make it easier for you to edit and revise it later. You should embed any diagrams, figures, or tables into the document. Be sure to write in a persuasive style; you are trying to convince others that your claim is acceptable or valid.

National Science Teachers Association

Checkout Questions

Lab 13. Characteristics of Minerals: What Are the Identities of the Unknown Minerals?

1. Why is it possible to use physical properties to identify minerals found in rocks?

2. An Earth scientist has a rock she wants to identify and a list of known mineral characteristics. The rock is black with gold metallic specks. Depending on where she runs the streak test, streaks appear black. Its Mohs hardness is 6.2, and its density is 5.2. A table of known mineral characteristics is below.

Mineral	Color	Streak	Luster	Mohs Hardness	Density
Halite	Colorless or white when pure; impurities produce any color but usually yellow, gray, black, brown, red	White	Vitreous	2.5	2
Magnetite	Black to silver gray	Black	Metallic to submetallic	5–6.5	5.2
Muscovite	Thick specimens often appear to be black, brown, or silver in color; however, when split into thin sheets muscovite is colorless, sometimes with a tint of brown, yellow, green, or rose	White, often sheds tiny flakes	Pearly to vitreous	2.5–3	2.8–2.9
Pyrite	Brass-yellow	Greenish black to brownish black	Metallic	6–6.5	4.9–5.2

a. Which minerals does the rock likely contain?

b. How do you know?

3. There is no universal step-by step scientific method that all scientists follow.

    a. I agree with this statement.

    b. I disagree with this statement.

Explain your answer, using an example from your investigation about characteristics of minerals.

4. "The rock's color is red" is an example of data.

   a. I agree with this statement.

   b. I disagree with this statement.

   Explain your answer, using an example from your investigation about characteristics of minerals.

5. Scientists often need to look for patterns that occur in the data they collect and analyze. Explain why identifying patterns are important, using an example from your investigation about characteristics of minerals.

6. In nature, the way something is structured often determines its function or places limits on what it can or cannot do. Explain why it is important to keep in mind the relationship between structure and function when attempting to collect or analyze data, using an example from your investigation about the characteristics of minerals.

Lab Handout

Lab 14. Distribution of Natural Resources: Which Proposal for a New Copper Mine Maximizes the Potential Benefits While Minimizing the Potential Costs?

Introduction

Copper is a useful metal because of its unique physical properties (see Figure L14.1). It is easily stretched, molded, and shaped; is resistant to corrosion; and is a good conductor of heat and electricity. People use copper to make the wires and pipes found in homes or businesses. In addition, manufacturing companies use copper to make the appliances (such as refrigerators and ovens) and consumer electronics (such as phones and computers) that we use in our homes every day. In addition, copper is an essential component in the motors, wiring, radiators, brakes, and bearings found in cars and trucks. The average car contains 1.5 kilometers (0.9 mile) of copper wire, and the total amount of copper in a vehicle can range from 20 kilograms (44 pounds) in a small car to 45 kilograms (99 pounds) in a large luxury or hybrid car (USGS 2009).

FIGURE L14.1 _____

A sample of native copper extracted from a mine

Copper is a natural resource. Unfortunately, it is only located in a few specific sites around the world; these sites are called deposits. Earth scientists classify copper deposits based on the geologic process that created them. *Porphyry copper deposits*, which contain about 60% of the world's copper, form when a large mass of molten rock cools and solidifies deep in the Earth's crust. These deposits are large and often contain between 100 million to 5 billion metric tons of copper-bearing rock called ore. The ore in these deposits, however, only contains 0.2%–1% copper by weight. Porphyry copper deposits are often located in East Asia and in the mountainous regions of western North and South America. *Sediment-hosted copper deposits*, in contrast, form due to the deposition and subsequent cementation of sediments. Sediment-hosted copper deposits contain about 20% of the world's copper. These deposits are smaller than porphyry deposits and usually only contain 1 million to 100 million metric tons of ore. The ore in these deposits, however, often contains 2%–6% copper by weight. People have found sediment-hosted copper deposits in Zambia, Zaire, Europe, central Asia, and in the north central United States.

Most copper deposits, regardless of how they formed, have a definable boundary. An important component of the U.S. Geological Survey's (USGS) Mineral Resources Program is to identify the location and boundaries of untouched copper deposits, estimate the amount of copper that is likely in each one, and then share this information with scientists, other government agencies, and mining companies.

Copper deposits are often located far beneath Earth's surface. People therefore rely on mining companies to access and refine the copper in a deposit. A mining project often lasts for decades and include three major tasks. The first task in a mining project is to remove the ore from the deposit. Before miners can access the ore in a deposit, they must first remove the rock and soil that covers it. Miners call this rock and soil overburden. A deposit that is deeper costs more in terms of time and resources to access than a deposit that is near the surface. Mining companies tend to use underground or open-pit mining methods to remove overburden and the ore (see Figure L14.2 for an example of an open-pit copper mine).

The second major task is to separate the copper-based minerals from the ore. This task is called milling. This task is important because ore contains very little copper by weight, so

FIGURE L14.2

An open-pit copper mine in Bisbee, Arizona

most of the minerals within the ore is considered waste by mining companies. To separate the copper-based minerals from the ore, mining companies must first crush and grind the ore into a powder. Mining companies then use a method called froth flotation to separate the copper-based minerals from minerals that do not contain copper. The result of the froth flotation process is a solution called copper concentrate, which is about 25%–35% copper by weight.

The third major task is to make a pure copper metal that is usable by manufacturers. Mining companies accomplish this task by first mixing the copper concentrate with silica and the heating the mixture to 1200°C. This process, which is called smelting, removes many impurities and produces a liquid of copper sulfide that is about 60% copper by weight. This liquid is called copper matte. The mining company then heats the copper matte to remove the sulfur and produce a material called blister copper, which is about 99% copper by weight. Next, the blister copper is refined and molded into sheets so it can go through a chemical process called electrolysis that removes the few remaining impurities in the metal. After the electrolysis is complete, the resulting copper metal is about 99.95% pure. The mining company can then sell the pure copper by the pound to various manufacturing companies.

Like the majority of human activities, accessing and refining copper produces waste materials. Waste is a general term for any material that currently has little or no economic value. There are different types of mine waste materials that vary in their physical and chemical composition, their potential for environmental contamination, and how mine companies manage them over time. There are five major categories of mine waste:

- *Overburden* is the excess soil and rock that is removed to gain access to the mineral-rich ore. Overburden consists of acid-generating and non-acid-generating rock. Water that flows over or through acid-generating rock becomes more acidic. This acidic water can enter a nearby stream, river, or lake and cause environmental contamination.

- *Tailings* is a water-based slurry that is produced when a mineral is separated from an ore. It consists of finely ground rock, mineral waste products, and toxic processing chemicals. Mine companies typically store tailings in large artificial ponds (see Figure L14.3, p. 168). However, some companies are considering dumping tailings into oceans as an alternative method for disposal.

- *Slag* is a by-product of smelting. It consists of iron oxide and silicon dioxide. Slag is nontoxic and can be used to make concrete, roads, the grit used in sandblasters, and blocks for buildings.

- *Atmospheric emissions* include dust and the sulfur oxides from the smelting process. Atmospheric emissions vary in their composition and potential for environmental contamination.

FIGURE L14.3

Aerial photograph of the Kennecott mine tailings storage pond near Salt Lake City, Utah

- *Mine water* is the water that is used to extract, crush, and grind ore and to control dust. It often contains dissolved minerals and metals or trace amounts of processing chemicals. Mine water can enter nearby streams, rivers, and lakes and cause environmental contamination.

Although many historic mining operations were not required to conduct their mining activities in ways that would reduce the negative impact on the environment, current federal and state regulations now require mining operations to use environmentally sound practices to minimize the effects of mineral development on human and ecosystem health. In this investigation, you will have an opportunity to examine several different proposals for developing and managing a copper mining operation to determine which one maximizes the benefits of mining copper while minimizing the potential costs. This is important to consider because the United States currently uses more copper (1.8 million metric tons in 2016) for construction and manufacturing than it produces (1.41 million tons in 2016). Although the United States can reduce its need for new copper mining projects by implementing more conservation, reuse, and recycling programs, it is also important to consider ways to minimize the impact of mining projects when new ones are needed.

Your Task

Use what you know about the uneven distribution of natural resources; the economic, social, and environmental costs associated with accessing natural resources; the importance of looking for patterns; and the nature of cause-and-effect relationships to identify the best proposal for starting a new copper mining project. Your assessment of each proposal must include an analysis of the potential benefits and costs associated with each proposal to determine which one has the highest benefit-to-cost ratio.

The guiding question of this investigation is, **Which proposal for a new copper mine *maximizes the potential benefits while minimizing the potential costs?***

Materials

You may use the following resources during your investigation:

- Google Earth is an interactive map available at *www.google.com/earth,* and Google Maps is an interactive map available at *https://maps.google.com.*

- The U.S. Geological Survey (USGS) Global Assessment of Undiscovered Copper Resources is available at *https://mrdata.usgs.gov/sir20105090z.* This interactive map and database includes information about copper deposits all over the world.

- The USGS Mineral Commodity Summaries is available at *https://minerals.usgs.gov/minerals/pubs/mcs.* This online publication includes information about the market price of copper.

- Surf Your Watershed is available at *https://cfpub.epa.gov/surf/locate/index.cfm.* This Environmental Protection Agency (EPA) database provides information about stream water quality in different watersheds.

- The U.S. Fish and Wildlife Service Environmental Conservation Online System (ECOS) is available at *https://ecos.fws.gov/ecp/report/table/critical-habitat.html.* This system includes an interactive map that shows the location of protected habitats for threatened and endangered species.

Safety Precautions

Follow all normal lab safety rules.

Investigation Proposal Required?   ☐ Yes   ☐ No

Getting Started

Your teacher will give you several different copper mine proposals to evaluate. These proposals include different plans for accessing and refining the copper from a specific deposit within the United States. Mining companies must submit a proposal, which outlines their overall plan for developing and managing the extraction of a natural resource such as copper, to state and federal agencies for approval when they want to open a new mine. The overall plan for the mining project must be approved at both the state and federal level before the mining company can start digging. The copper mine proposals that you use during this investigation include the following information:

- Location of the mine
- Mining method
- Estimated life span of the mine
- Estimated number of new jobs that will be created when the mine opens
- Size of the site
- Total amount of ore (rock with copper in it) to be removed from the site
- Ore extraction rate
- Amount of waste that will be produced
- Waste management plan
- Expenses associated with opening, operating, and closing the mine

You will need to conduct a benefit-and-cost analysis of each proposal to determine which mine plan is the best one. A benefit-and-cost analysis requires the identification of the potential benefits of starting a new mining project and all the potential costs associated with opening, operating, and the closing the mine. Potential benefits associated with starting a new copper mine include such things as how much copper is in the deposit, how easy the deposit will be to access, how much ore that can be removed from the deposit, and the overall value of the copper on the open market. A mine can also have a positive impact on the local economy because it can create new jobs for people who live in that area. The potential costs, in contrast, include the amount of money needed to open, operate, and then close the mine, the negative impacts of mine waste on ecosystem health, and how the mine will be viewed by people who live in the area. You will therefore need to consider all of these issues, and potentially several others, as you conduct a benefit-and-cost analysis of each proposal during this investigation.

The first step in your benefit-and-cost analysis is to determine the location of each mine. To accomplish this step, you can enter the coordinates included with each proposal into Google Earth and/or Google Maps.

The second step in your benefit-and-cost analysis is to learn more about the copper deposit at a proposed location. You will need to use the USGS Global Assessment of Undiscovered Copper Resources database to accomplish this step. To use this database, simply zoom in on the location of the proposed mine on the interactive map. You can then click on the different deposits marked on the map until you find the name of the deposit you are interested in learning more about. You can then click on the name of the deposit to bring up information about it. As you use this resource, be sure to think about the following questions:

- What information will help you determine the potential benefits and costs of mining at this location?
- What would make one deposit more valuable than another deposit?
- How can you use mathematics to determine how much copper the mine could potentially produce based on the proposed amount of ore (rock with copper in it) that will be removed from the site?

The third step in your benefit-and-cost analysis is to determine the overall value of a proposed mine or the amount of revenue a mining company could potentially generate given the overall plan for a proposed mine. This is important to consider because a mine must be profitable to stay open. A mining company, in other words, must generate more money from selling the copper that it extracts from the mine than it spends to open, operate, and then close the mine. To determine how much money a mining company could make by selling the copper that it extracts from a mine, you can use the USGS Mineral Commodity Summaries. As you use this resource, be sure to think about the following questions:

- What information will help you determine the potential value of the copper in a deposit?
- How can you use mathematics to determine how much money a mining company could generate per day based on each copper mine proposal?
- What information will help you determine the potential cost of operating a proposed mine?
- How can you use mathematics to determine if a proposed mine will be profitable or not?

The fourth step in your benefit-and-cost analysis is to investigate the potential environmental impacts of each mine proposal. Mines can produce waste that can pollute water, contaminate soil, and destroy habitats. To determine potential environmental impacts of opening and operating mines, you will need to identify any streams, lakes, and important wildlife habitats around proposed mine sites. You can determine which streams and lakes are located near the proposed mine site by using Google Earth and/or Google Maps. You can also check the current water quality of these streams and lakes by accessing the Surf Your Watershed database. You can determine if there are any protected habitats of threatened or endangered species near a proposed mine by accessing the U.S. Fish and Wildlife Service Environmental Conservation Online System. As you use these resources, be sure to think about the following questions:

LAB 14

- What information will help you determine the potential environmental impacts of opening, operating, and closing a mine?

- What measurement scale or scales should you use to collect data?

- How can you describe or quantify the potential environmental impact of each proposed mine?

- What cause-and-effect relationships will you need to keep in mind as you use these resources?

The final step in your benefit-and-cost analysis is to choose between the different proposals based on the strengths and weakness of each one. To evaluate the trade-offs in each proposal fairly, you may want to create a decision matrix. A decision matrix includes rows for each proposal and columns that provide different evaluation criteria. Table L14.1 is an example of a decision matrix. Evaluation criteria might include such things as potential value of extracted copper, amount of waste produced, waste management plan, and potential habitat loss. You will need to determine which evaluation criteria to use based on what you know about the uneven distribution of natural resources and the economic, social, and environmental costs that are often associated with accessing natural resources. Once you have determined the evaluation criteria that your group will use, rank order each proposal on each criterion. Give the best option the highest number (which is a 4 if there are four different mining proposals) and the worst option a 1. You can then total the rankings, and the proposal with the highest overall score is the best one.

TABLE L14.1

An example of a decision matrix

Proposal	Evaluation criteria					Overall score
	1	2	3	4	5	
A						
B						
C						
D						

Connections to the Nature of Scientific Knowledge and Scientific Inquiry

As you work through your investigation, be sure to think about

- how scientific knowledge changes over time, and

- the types of questions that scientists can investigate.

Initial Argument

Once your group has finished collecting and analyzing your data, your group will need to develop an initial argument. Your initial argument needs to include a claim, evidence supporting your claim, and a justification of the evidence. The *claim* is your group's answer to the guiding question. The *evidence* is an analysis and interpretation of your data. Finally, the *justification* of the evidence is why your group thinks the evidence matters. The justification of the evidence is important because scientists can use different kinds of evidence to support their claims. Your group will create your initial argument on a whiteboard. Your whiteboard should include all the information shown in Figure L14.4.

FIGURE L14.4 _____

Argument presentation on a whiteboard

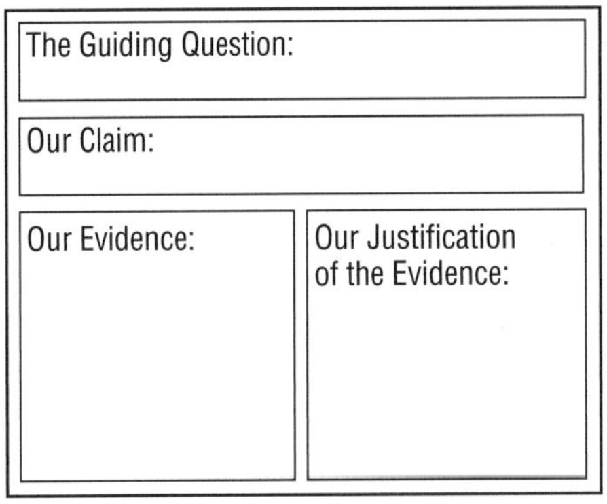

The Guiding Question:	
Our Claim:	
Our Evidence:	Our Justification of the Evidence:

Argumentation Session

The argumentation session allows all of the groups to share their arguments. One or two members of each group will stay at the lab station to share that group's argument, while the other members of the group go to the other lab stations to listen to and critique the other arguments. This is similar to what scientists do when they propose, support, evaluate, and refine new ideas during a poster session at a conference. If you are presenting your group's argument, your goal is to share your ideas and answer questions. You should also keep a record of the critiques and suggestions made by your classmates so you can use this feedback to make your initial argument stronger. You can keep track of specific critiques and suggestions for improvement that your classmates mention in the space below.

Critiques of our initial argument and suggestions for improvement:

If you are critiquing your classmates' arguments, your goal is to look for mistakes in their arguments and offer suggestions for improvement so these mistakes can be fixed. You should look for ways to make your initial argument stronger by looking for things that the other groups did well. You can keep track of interesting ideas that you see and hear during the argumentation in the space below. You can also use this space to keep track of any questions that you will need to discuss with your team.

Interesting ideas from other groups or questions to take back to my group:

Once the argumentation session is complete, you will have a chance to meet with your group and revise your initial argument. Your group might need to gather more data or design a way to test one or more alternative claims as part of this process. Remember, your goal at this stage of the investigation is to develop the best argument possible.

Report

Once you have completed your research, you will need to prepare an *investigation report* that consists of three sections. Each section should provide an answer for the following questions:

1. What question were you trying to answer and why?

2. What did you do to answer your question and why?

3. What is your argument?

Your report should answer these questions in two pages or less. You should write your report using a word processing application (such as Word, Pages, or Google Docs), if possible, to make it easier for you to edit and revise it later. You should embed any diagrams, figures, or tables into the document. Be sure to write in a persuasive style; you are trying to convince others that your claim is acceptable or valid.

Reference

U.S. Geological Survey (USGS). 2009. Copper—A metal for the ages. U.S. Department of the Interior, USGS. Available online at *https://pubs.usgs.gov/fs/2009/3031/FS2009-3031.pdf*.

LAB 14

Lab 14. Distribution of Natural Resources: Which Proposal for a New Copper Mine Maximizes the Potential Benefits While Minimizing the Potential Costs?

1. The map below shows the locations of the active mines producing copper in the United States as of 2003.

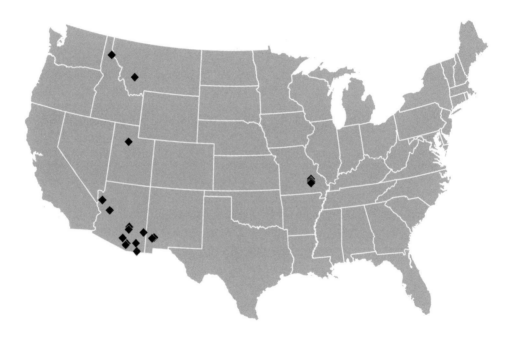

    a. Why are most copper mines located in the same region of the United States?

b. What are some of the potential benefits of opening a new copper mine?

c. What are some of the potential costs of opening a new copper mine?

2. Once something has been discovered in science, it is proven and will never change.

a. I agree with this statement.

b. I disagree with this statement.

Explain your answer, using an example from your investigation about the distribution of natural resources or other investigations you have done.

3. Scientists can only investigate certain types of questions.

    a. I agree with this statement.

    b. I disagree with this statement.

Explain your answer, using an example from your investigation about the distribution of natural resources or other investigations you have done.

4. Identifying patterns is one of the important tasks that scientists carry out. Give an example of a pattern and why it was important for scientists to identify this pattern using information from your investigation about the distribution of natural resources.

5. Natural phenomena have causes, and uncovering causal relationships is a major activity of science. Explain why it is important to uncover causal relationships, using an example from your investigation about the distribution of natural resources.

SECTION 5
Weather and Climate

Introduction Labs

LAB 15

Lab Handout

Lab 15. Air Masses and Weather Conditions: How Do the Motions and Interactions of Air Masses Result in Changes in Weather Conditions?

Introduction

Meteorology is the study of the atmosphere. Meteorologists study the atmosphere so they can make accurate predictions about future weather conditions. In fact, meteorologists have generated detailed weather maps that include information about the atmosphere and current weather conditions in different regions of the United States for over a century. In the late 1800s, for example, newspapers printed a weather map every morning. An example of a U.S. daily weather map from 1899 can be seen in Figure L15.1. People relied on these weather maps to make predictions about the weather so they could make better decisions about what to do during the day, where to go, or what to wear. As technologies advanced, we developed faster ways to deliver up-to-date information about current weather conditions, including radios, live television broadcasts, and internet posts. Meteorologists are now able to consult many sources, such as computer models, real-time weather station data, and Doppler radar, to generate forecasts. These forecasts are often very accurate and can predict general daily weather up to 10 days in advance.

FIGURE L15.1

The U.S. Daily Weather Map for February 8, 1899

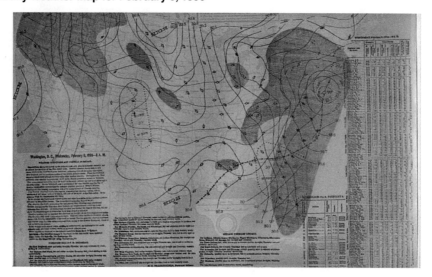

Although weather maps may look complicated, they simply display information about air masses and other atmospheric conditions. An *air mass* is a large body of air that has a relatively uniform temperature and humidity level. The curved lines on a weather map mark the boundaries of different air masses. The center of an air mass is marked with the letter *H* or *L*, which denotes whether the air mass has a high or low atmospheric pressure. The temperature of air affects the atmospheric pressure within an air mass. Warm air consists of molecules that are moving faster and more spread out compared with cold air. This makes hot air less dense than cold air.

When thinking about air masses, it is important to remember that air masses are three-dimensional; they spread out across a region (North, East, South, and West), while also extending upward from the surface of Earth. When an air mass warms at Earth's surface, it becomes less dense and begins to rise. As it rises farther away from Earth's surface, it cools, becomes denser, and sinks. When two air masses meet, the warmer, less dense air mass will rise above the colder, denser air mass.

The area where two air masses meet is called a front. Meteorologists categorize fronts based on the nature of the air mass that is moving into an area or how two or more air masses are interacting with each other. A cold front, for example, refers to instances when a cold air mass moves into an area that was previously occupied by a warm air mass. On a weather map, lines with shapes on them represent different types of fronts. A line with triangles is used to indicate the boundary and movement of a cold front (see Figure L15.2). A line with semicircles is used to indicate the boundary and movement of a warm front (see Figure L15.3). The shapes always point in the direction an air mass is moving. A third type of front is called a stationary front. Stationary fronts form in areas where warm air masses and cold air masses move past each other in opposite directions (see Figure L15.4, p. 184). The warm air mass is always on the side of the line without the semicircles, and the cold air mass is always on the side of the line without the triangles.

The interaction between two air masses can cause a change in weather conditions. Meteorologists therefore

FIGURE L15.2

Cold fronts are shown with triangles on weather maps; the triangles point to the direction that the air is moving

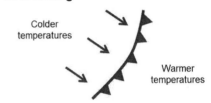
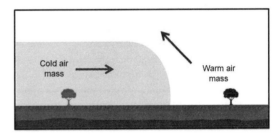

FIGURE L15.3

Warm fronts are shown with semicircles on weather maps; the semicircles indicate the direction that the air is moving

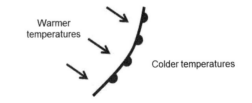
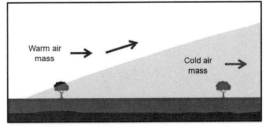

Note: A full-color version of Figures L15.2 and L15.3 can be downloaded from the book's Extras page at *www.nsta.org/adi-ess*.

LAB 15

Stationary fronts are depicted with triangles and semicircles on weather maps

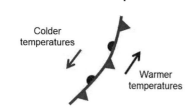

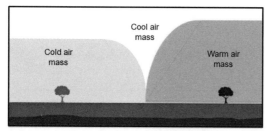

Note: A full-color version of this figure can be downloaded from the book's Extras page at *www.nsta.org/adi ess*.

track the movement of air masses to make predictions about future weather conditions. In this investigation, you will have an opportunity to use historical weather maps and weather data from several different regions to learn how the motions and complex interactions of air masses are related to changes in weather conditions. Your goal is to develop a conceptual model that you can use to not only explain how the motions and interactions of air masses result in specific weather conditions but also predict how weather conditions will change over time in a given area.

Your Task

Develop a conceptual model that can be used to explain weather conditions based on the movement and interactions of air masses. Your conceptual model must be based on what we know about the weather, the kinds of air masses found in the atmosphere, the importance of looking for patterns in nature, and cause-and-effect relationships. Once you have developed your model, you will need to test it to see if you can use it to make accurate predictions about the weather conditions in different cities on a given date.

The guiding question of this investigation is, *How do the motions and interactions of air masses result in changes in weather conditions?*

Materials

You may use any of the following materials during your investigation:

Equipment
- Computer or tablet with internet access

Other Resources
- Weather Map A (use to test your model)
- Weather Conditions Table A (use to test your model)
- Weather Map B (use to test your model)
- Weather Conditions Table B (use to test your model)

Safety Precautions

Follow all normal lab safety rules.

Investigation Proposal Required? ☐ Yes ☐ No

Getting Started

The first step in this investigation is to analyze an existing data set to determine how the movement or interaction of different kinds of air masses is related to specific weather conditions. To accomplish this goal, you will need to examine several different historical weather maps and look for patterns that you can use to explain and predict changes in weather conditions. You can access U.S. Daily Weather Maps from the National Oceanic and Atmospheric Administration (NOAA) / National Weather Service Weather Prediction Center at *www.wpc.ncep.noaa.gov/dwm/dwm.shtml.*

Once you have identified patterns in the historical weather maps, you can develop your conceptual model. A conceptual model is an idea or set of ideas that explains what causes a particular phenomenon in nature. People often use words, images, and arrows to describe a conceptual model. Your conceptual model needs to be able to explain changes in weather conditions based on the movement and interactions of air masses. The model also needs to be consistent with what we know about what causes changes in atmospheric pressure, the nature of wind, and the cycling of water on Earth.

The last step in this investigation is to test your model. To accomplish this goal, you can make predictions about the weather conditions at several different cities using the information found on Weather Maps A and B. Your teacher will identify the cities that you will need to include in your predictions. You can then use Weather Conditions Tables A and B to determine if your predictions were accurate. If you are able to use your model to make accurate predictions about the weather conditions in different cities, then you will be able to generate the evidence you need to convince others that the conceptual model you developed is valid or acceptable.

Connections to the Nature of Scientific Knowledge and Scientific Inquiry

As you work through your investigation, be sure to think about

- the difference between observations and inferences in science, and
- how scientists use different methods to answer different types of questions.

Initial Argument

Once your group has finished collecting and analyzing your data, your group will need to develop an initial argument. Your initial argument needs to include a claim, evidence to support your claim, and a justification of the evidence. The *claim* is your group's answer to the guiding question. The *evidence* is an analysis and interpretation of your data. Finally, the *justification* of the evidence is why your group thinks the evidence matters. The justification of the evidence is important because scientists can use different kinds of evidence to support their claims. Your group will create your initial argument on a whiteboard. Your

LAB 15

Argument presentation on a whiteboard

The Guiding Question:	
Our Claim:	
Our Evidence:	Our Justification of the Evidence:

whiteboard should include all the information shown in Figure L15.5.

Argumentation Session

The argumentation session allows all of the groups to share their arguments. One or two members of each group will stay at the lab station to share that group's argument, while the other members of the group go to the other lab stations to listen to and critique the other arguments. This is similar to what scientists do when they propose, support, evaluate, and refine new ideas during a poster session at a conference. If you are presenting your group's argument, your goal is to share your ideas and answer questions. You should also keep a record of the critiques and suggestions made by your classmates so you can use this feedback to make your initial argument stronger. You can keep track of specific critiques and suggestions for improvement that your classmates mention in the space below.

Critiques of our initial argument and suggestions for improvement:

If you are critiquing your classmates' arguments, your goal is to look for mistakes in their arguments and offer suggestions for improvement so these mistakes can be fixed. You should look for ways to make your initial argument stronger by looking for things that the other groups did well. You can keep track of interesting ideas that you see and hear during the argumentation in the space below. You can also use this space to keep track of any questions that you will need to discuss with your team.

Interesting ideas from other groups or questions to take back to my group:

Once the argumentation session is complete, you will have a chance to meet with your group and revise your initial argument. Your group might need to gather more data or design a way to test one or more alternative claims as part of this process. Remember, your goal at this stage of the investigation is to develop the best argument possible.

Report

Once you have completed your research, you will need to prepare an *investigation report* that consists of three sections. Each section should provide an answer for the following questions:

1. What question were you trying to answer and why?

2. What did you do to answer your question and why?

3. What is your argument?

Your report should answer these questions in two pages or less. You should write your report using a word processing application (such as Word, Pages, or Google Docs), if possible, to make it easier for you to edit and revise it later. You should embed any diagrams, figures, or tables into the document. Be sure to write in a persuasive style; you are trying to convince others that your claim is acceptable or valid.

Checkout Questions

Lab 15. Air Masses and Weather Conditions: How Do the Motions and Interactions of Air Masses Result in Changes in Weather Conditions?

1. What are air masses and how do meteorologists classify them?

2. A meteorologist is given the weather map below and is asked to predict the upcoming weather for the four cities marked with the letters A, B, C, and D in black circles.

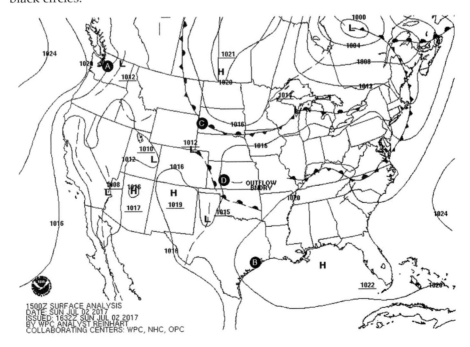

    a. Which city or cities will probably experience rain in the near future?

b. How do you know?

3. The statement "an air mass with low pressure is replacing one with high pressure" is an example of an observation.

   a. I agree with this statement.

   b. I disagree with this statement.

   Explain your answer, using an example from your investigation about air masses and weather conditions.

4. Scientists, regardless of their discipline, follow the same step-by step method to answer questions about natural phenomena.

   a. I agree with this statement.

   b. I disagree with this statement.

   Explain your answer, using an example from your investigation about air masses and weather conditions.

5. Scientists often need to look for patterns during an investigation. Explain why identifying patterns is important in science, using an example from your investigation about air masses and weather conditions.

6. Scientists often attempt to uncover a cause-and-effect relationship as part of an investigation. Explain what a cause-and-effect relationship is and why these relationships are so important in science, using an example from your investigation about air masses and weather conditions.

LAB 16

Lab Handout

Lab 16. Surface Materials and Temperature Change: How Does the Nature of the Surface Material Covering a Specific Location Affect Heating and Cooling Rates at That Location?

Introduction

Inner cities and suburbs tend to be much warmer than rural areas as a result of land use and human activities. Scientists call this phenomenon the urban heat island effect. Take Atlanta, Georgia, as an example. Figure L16.1 shows two Landsat satellite images of Atlanta taken on September 28, 2000. Image A is a true-color picture of Atlanta, where trees and other vegetation are dark green and roads or buildings are different shades of gray. Image B, in contrast, is a map of land surface temperature. In image B, cooler temperatures are yellow and hotter temperatures are red. Downtown Atlanta is in the center of both images. On this day in 2000, the temperature in the areas in and around downtown Atlanta reached 30°C (86°F), while some of the less densely developed areas outside of the city only reached 20°C (68°F). Las Vegas, Nevada, also experiences a significant urban heat island effect. On hot summer days, it can be 13°C (24°F) warmer in downtown Las Vegas than it is in the surrounding desert. Downtown Las Vegas also has, on average, 22 more days that are above 32°C (90°F) each year when compared with the surrounding rural areas. Most major cities in the United States, including Dallas, Phoenix, New York, Los Angeles, Denver, and Washington, D.C., experience a significant urban heat island effect.

FIGURE L16.1

Two Landsat satellite images of Atlanta: (a) a true-color picture of the city and (b) an image showing the differences in temperature for the city and areas around the city in the afternoon

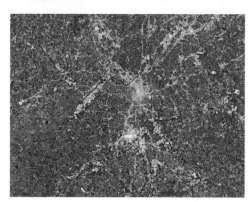 

a   b

Note: A full-color version of this figure can be downloaded from the book's Extras page at www.nsta.org/adi-ess.

Surface Materials and Temperature Change

How Does the Nature of the Surface Material Covering a Specific Location Affect Heating and Cooling Rates at That Location?

In our everyday conversations, we often use the terms *temperature* and *heat* interchangeably. In science, however, these two terms have different meanings. *Temperature* is used to describe the average kinetic energy of the atoms or molecules that make up an object. *Heat*, on the other hand, is the transfer of thermal energy into, within, or out of an object. There are three ways thermal energy can transfer into, within, or out of an object: conduction, convection, and radiation.

- *Conduction* is the transfer of thermal energy due to the collision of the atoms or molecules within an object or between two objects in contact. Thermal energy always transfers from an object or area of higher temperature to an object or area of lower temperature.

- *Convection* is the transfer of thermal energy due to the mass movement or circulation of particles within a fluid. Fluids are liquids (like lakes or oceans) and gases (such as air).

- *Radiation* is the transfer of thermal energy through electromagnetic waves. An example of radiation is what happens to a car when it sits in the sunlight on a hot summer day. The car absorbs sunlight, and the temperature of the car increases. As more sunlight is absorbed over time, the temperature of the car increases as well.

When thermal energy transfers into an object, the temperature of the object will increase; when heat transfers out of an object, the temperature will decrease. However, not all objects will undergo the same change in temperature when the same amount of thermal energy is added to them. For example, adding 1 joule (J) of thermal energy to a 1-kilogram (kg) sample of lead will cause the piece of lead to increase in temperature by about 8°C. Adding 1 J of thermal energy to a 1 kg sample of water, however, will only increase the temperature of the water by approximately 0.2°C.

There are many potential explanations for the urban heat island effect. First, cities have more people living in them. Some scientists have therefore speculated that a higher concentration of people using air conditioners may be causing the urban heat island effect because air conditioners remove heat from the air in buildings and transfer it outside. Other scientists have suggested that car exhaust is causing the urban heat island effect because the gases in exhaust can trap thermal energy, and there are many more cars in cities than there are in rural areas. Finally, other scientists suggest that the materials we use to build roads, homes, and other buildings are the source of the urban heat island effect. Scientists studying this possibility note that in cities that experience an urban heat island effect, there tends to be a much higher concentration of concrete inside the city when compared with the area surrounding the city, which tends to be covered by naturally occurring materials such as plants, sand, or water.

Your Task

Use what you know about heat and temperature, cause-and-effect relationships, and stability and change in systems to plan and carry out an investigation that will allow you to determine the relationship between the materials covering an area and the rate at which the temperature of that area changes over time. This investigation will aid you in understanding the underlying cause of urban heat islands.

The guiding question of this investigation is, *How does the nature of the surface material covering a specific location affect heating and cooling rates at that location?*

Materials

You may use any of the following materials during your investigation:

Consumables	Equipment	
• Water	• Safety glasses or goggles (required)	• Partial immersion (nonmercury) thermometers
• Soil	• Chemical-resistant apron (required)	• Digital or laser thermometer (optional)
• Dark sand	• Gloves (required)	
• Light sand	• Styrofoam cups	• Graduated cylinder (250 ml)
• Concrete	• Electronic or triple beam balance	• Support stand
• Sod	• Infrared lamp and reflector	• Ruler

Safety Precautions

Follow all normal lab safety rules. In addition, take the following safety precautions:

- Wear sanitized indirectly vented chemical-splash goggles and chemical-resistant, nonlatex aprons, and gloves throughout the entire investigation (which includes setup and cleanup).

- Use only a GFCI-protected electrical receptacle for the lamp to prevent or reduce risk of shock.

- Handle the infrared lamp with care; it can get hot enough to burn skin.

- Do not spill or splash water on the hot lamp bulb—this can crack glass and form a projectile.

- Report and clean up spills immediately, and avoid walking in areas where water has been spilled.

- Wash hands with soap and water when done collecting the data and after completing the lab.

Investigation Proposal Required?   ☐ Yes   ☐ No

Surface Materials and Temperature Change

How Does the Nature of the Surface Material Covering a Specific Location Affect Heating and Cooling Rates at That Location?

Getting Started

To answer the guiding question, you will need to design and carry out an experiment. Figure L16.2 shows how you can use a heat lamp to warm different types of materials, such as soil, water, sand, concrete, or sod (grass). Before you begin to design your experiment using this equipment, think about what type of data you need to collect, how you will collect the data, and how you will analyze the data.

FIGURE L16.2

How to use a heat lamp to warm different types of materials

To determine *what type of data you need to collect,* think about the following questions:

- What are the components of the system you are studying?
- Which factor(s) might control the rate of change in this system?
- How will you measure how quickly the materials heat up (or rate of change)?
- I Iow will you measure how quickly the materials cool down (or rate of change)?

To determine *how you will collect the data,* think about the following questions:

- What conditions need to be satisfied to establish a cause-and-effect relationship?
- What will serve as your independent variable and dependent variables?
- How will you vary the independent variable while holding the other variables constant?
- How will you make sure the amount of each material is the same?
- How will you make sure that your data are of high quality (i.e., how will you reduce error)?
- How will you keep track of and organize the data you collect?

To determine *how you will analyze the data,* think about the following questions:

- What type of calculations will you need to make?
- How could you use mathematics to document a difference between conditions?
- What type of table or graph could you create to help make sense of your data?
- How will you determine if rates of change are the same or different?

LAB 16

Connections to the Nature of Scientific Knowledge and Scientific Inquiry

As you work through your investigation, be sure to think about

- the difference between laws and theories in science, and
- the types of questions that scientists can investigate.

Initial Argument

Once your group has finished collecting and analyzing your data, your group will need to develop an initial argument. Your initial argument needs to include a claim, evidence to support your claim, and a justification of the evidence. The *claim* is your group's answer to the guiding question. The *evidence* is an analysis and interpretation of your data. Finally, the *justification* of the evidence is why your group thinks the evidence matters. The justification of the evidence is important because scientists can use different kinds of evidence to support their claims. Your group will create your initial argument on a whiteboard. Your whiteboard should include all the information shown in Figure L16.3.

FIGURE L16.3 _____

Argument presentation on a whiteboard

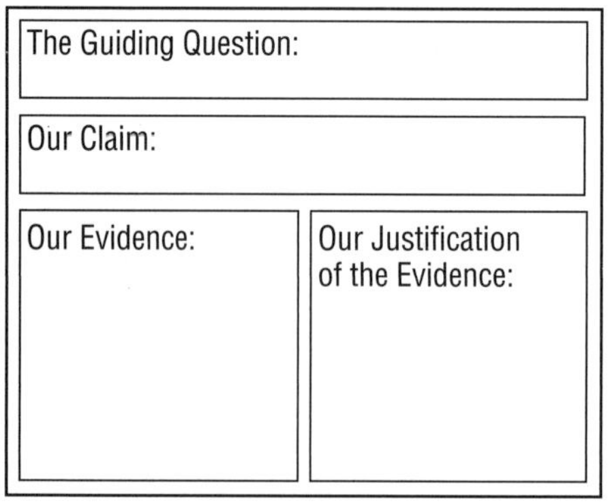

The Guiding Question:	
Our Claim:	
Our Evidence:	Our Justification of the Evidence:

Argumentation Session

The argumentation session allows all of the groups to share their arguments. One or two members of each group will stay at the lab station to share that group's argument, while the other members of the group go to the other lab stations to listen to and critique the other arguments. This is similar to what scientists do when they propose, support, evaluate, and refine new ideas during a poster session at a conference. If you are presenting your group's argument, your goal is to share your ideas and answer questions. You should also keep a record of the critiques and suggestions made by your classmates so you can use this feedback to make your initial argument stronger. You can keep track of specific critiques and suggestions for improvement that your classmates mention in the space below.

Critiques of our initial argument and suggestions for improvement:

If you are critiquing your classmates' arguments, your goal is to look for mistakes in their' arguments and offer suggestions for improvement so these mistakes can be fixed. You should look for ways to make your initial argument stronger by looking for things that the other groups did well. You can keep track of interesting ideas that you see and hear during the argumentation in the space below. You can also use this space to keep track of any questions that you will need to discuss with your team.

Interesting ideas from other groups or questions to take back to my group:

LAB 16

Once the argumentation session is complete, you will have a chance to meet with your group and revise your initial argument. Your group might need to gather more data or design a way to test one or more alternative claims as part of this process. Remember, your goal at this stage of the investigation is to develop the best argument possible.

Report

Once you have completed your research, you will need to prepare an *investigation report* that consists of three sections. Each section should provide an answer for the following questions:

1. What question were you trying to answer and why?

2. What did you do to answer your question and why?

3. What is your argument?

Your report should answer these questions in two pages or less. You should write your report using a word processing application (such as Word, Pages, or Google Docs), if possible, to make it easier for you to edit and revise it later. You should embed any diagrams, figures, or tables into the document. Be sure to write in a persuasive style; you are trying to convince others that your claim is acceptable or valid.

National Science Teachers Association

Checkout Questions

Lab 16. Surface Materials and Temperature Change: How Does the Nature of the Surface Material Covering a Specific Location Affect Heating and Cooling Rates at That Location?

1. If all objects are in the sunlight for the same time, why do some objects increase in temperature more than others?

2. Using data from your lab, explain how the design of cities contributes to heat islands. Make a recommendation for an urban planner on how to reduce the degree to which a city is a heat island.

LAB 16

3. An experiment is one possible method for answering a question in science. In an experiment, scientists develop a systematic plan for recording data. They do not manipulate or change any variables in order to answer their questions.

    a. I agree with this description of an experiment.

    b. I disagree with this description of an experiment.

Explain your answer, using an example from your investigation about surface materials and temperature change.

4. Theories and laws are both important in science. Theories provide explanations for why phenomena occur, and laws provide descriptions of phenomena.

    a. I agree with this statement.

    b. I disagree with this statement.

Explain your answer, using an example from your investigation about surface materials and temperature change.

5. Natural phenomena have causes, and uncovering causal relationships is a major activity of science. Explain what a causal relationship is and why it important to identify causal relationships in science, using an example from your investigation about surface materials and temperature change.

6. In science, it is important to understand what factors influence rates of change in a system. Explain why this is so important, using an example from your investigation about surface materials and temperature change.

Lab Handout

Lab 17. Factors That Affect Global Temperature: How Do Cloud Cover and Greenhouse Gas Concentration in the Atmosphere Affect the Surface Temperature of Earth?

Introduction

All matter in the universe radiates energy across a range of wavelengths in the electromagnetic spectrum. Hotter objects tend to emit radiation with shorter wavelengths than cooler objects. The hottest objects in the universe, as a result, mostly emit gamma rays and x-rays. Cooler objects, in contrast, emit mostly longer-wavelength radiation, including visible light, infrared (IR), microwaves, and radio waves. The surface of the Sun has a temperature of about 5500°C or about 10000°F. At that temperature, most of the energy the Sun radiates is visible and near-IR light.

When sunlight first reaches Earth, some of it is reflected back out into space and some is absorbed by the atmosphere. The rest of the sunlight travels through the atmosphere and then hits the surface of Earth. The energy from the sunlight is absorbed by the surface and warms it. All objects, including Earth's surface, emit (or give off) IR radiation. The hotter an object is, the more IR radiation it emits. The amount of IR radiation emitted by Earth's surface therefore increases as it warms. The atmosphere traps some of this IR radiation before it can escape into space. The trapped IR radiation in the atmosphere helps keep the temperature of Earth warmer than it would be without the atmosphere. Scientists call the warming of the atmosphere that is caused by trapped IR radiation the *greenhouse effect* (see Figure L17.1). Many gases that are found naturally in Earth's atmosphere, including water vapor, carbon dioxide, methane, nitrous oxide, and ozone, are called greenhouse gases because these gases are able to trap IR energy in the atmosphere.

The amount of energy that enters and leaves the Earth system is directly related to the average global temperature of the Earth. The Earth system, which includes the surface and the atmosphere, currently absorbs an average of about 340 watts of solar power per square meter over the course of the year (NASA n.d.). The Earth system also emits about the same amount of IR energy into space. The average global surface temperature, as a result, tends to be stable over time. However, if something were to change the amount of energy that enters or leaves this system, then the flow of energy would be unbalanced and the average global temperature would change in response. Therefore, any change to the Earth system that affects how much energy enters or leaves the system can cause a significant change in Earth's average global temperature.

The average global surface temperature of Earth has increased approximately 0.8°C (1.4°F) over the last 100 years (NASA Goddard Institute for Space Studies 2016). There are at least two potential explanations for this observation. One explanation is that the

FIGURE L17.1

The greenhouse effect

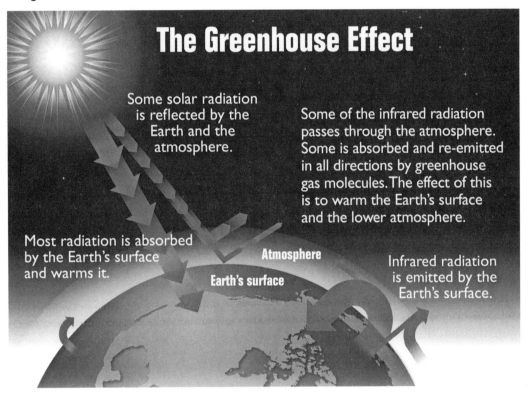

Note: A full-color version of this figure can be downloaded from the book's Extras page at *www.nsta.org/adi-ess.*

average global surface temperature of Earth normally increases and decreases over time and the current increase in temperature is just a normal part of this cycle. These changes could be to due to differences in the Sun's brightness, Milankovitch cycles (small variations in the shape of Earth's orbit and its axis of rotation that occur over thousands of years), or an increase or decrease in cloud cover (clouds form when water vapor in the air condenses into water droplets or ice). This explanation, however, does not account for the rapid increase in the average global surface temperature. An alternative explanation, which is the consensus view of the scientific community, is that humans have caused the rapid increase in the average global surface temperature of Earth by adding large amounts of greenhouse gases to the atmosphere. The addition of greenhouse gases magnifies the greenhouse effect. The atmosphere, as a result, traps more IR radiation and emits less IR energy out into space.

Before you can evaluate the merits of these two explanations for the observed change in average global surface temperature, it is important for you to understand how energy from the Sun interacts with the surface of the Earth and the various components of the atmosphere, such as clouds. You will therefore need to learn more about the relationships between surface temperature, cloud cover, and greenhouse gas levels in the atmosphere.

Your Task

Use a computer simulation and what you know about stability and change and the importance of tracking how energy flows into, within, and out of systems to determine how the temperature of Earth responds to changes in the amount of cloud cover and the concentration of carbon dioxide in the atmosphere.

The guiding question of this investigation is, *How do cloud cover and greenhouse gas concentration in the atmosphere affect the surface temperature of Earth?*

Materials

You will use an online simulation called *The Greenhouse Effect* to conduct your investigation; the simulation is available at *https://phet.colorado.edu/en/simulation/legacy/greenhouse*.

Safety Precautions

Follow all normal lab safety rules.

Investigation Proposal Required?   ☐ Yes   ☐ No

Getting Started

The *Greenhouse Effect* computer simulation models how energy flows into, within, and out of the Earth system and records changes in average global temperature over time (see Figure L17.2). It shows the surface of Earth as a green strip. Above the green strip there is a blue atmosphere and black space at the top. Yellow dots stream downward representing photons of sunlight. Red dots represent photons of IR light that are emitted by the surface of Earth and travel toward space. The greenhouse gas concentration in the atmosphere, including amounts of water vapor (H_2O), carbon dioxide (CO_2), methane (CH_4), and nitrous oxide (N_2O), can be changed so it reflects the current level of these gases, the level in 1750, the level during the last ice age, or a level of your choice. Clouds can also be added or removed from the atmosphere. Greenhouse gases block IR light (energy) that is emitted by Earth's surface. Clouds can block sunlight and IR photons.

To answer the research question, you must determine what type of data you need to collect, how you will collect it, and how will you analyze it. To determine *what type of data you need to collect*, think about the following questions:

- What are the boundaries and components of the system you are studying?
- How do the components of the system interact with each other?
- When is this system stable, and under which conditions does it change?

Factors That Affect Global Temperature
How Do Cloud Cover and Greenhouse Gas Concentration in the Atmosphere Affect
the Surface Temperature of Earth?

FIGURE L17.2

A screenshot from *The Greenhouse Effect* simulation

Note: A full-color version of this figure can be downloaded from the book's Extras page at *www.nsta.org/adi-ess.*

- Which factor(s) might control the rate of change in this system?
- How can you describe the components of the system quantitatively?
- How could you keep track of changes in this system quantitatively?
- How can you track how energy flows into, out of, or within this system?

To determine *how you will collect your data,* think about the following questions:

- What type of measurements or observations will you need to record during your investigation?
- How often will you need to make these measurements or observations?
- What will serve as your dependent variable?
- What will serve as a control condition?
- What types of treatment conditions will you need to set up?
- How many trials will you need to run in each condition?

- How long will you let the simulation run before you collect data?
- How will you keep track of and organize the data you collect?

To determine *how you will analyze the data,* think about the following questions:

- What types of patterns might you look for as you analyze your data?
- How could you use mathematics to describe a change over time?
- How could you use mathematics to document a difference between treatment and control conditions?
- What type of calculations will you need to make?
- What type of graph could you create to help make sense of your data?

Connections to the Nature of Scientific Knowledge and Scientific Inquiry

As you work through your investigation, be sure to think about

- how scientific knowledge can change over time, and
- the types of questions that scientists can investigate.

Initial Argument

Once your group has finished collecting and analyzing your data, your group will need to develop an initial argument. Your initial argument needs to include a claim, evidence to support your claim, and a justification of the evidence. The *claim* is your group's answer to the guiding question. The *evidence* is an analysis and interpretation of your data. Finally, the *justification* of the evidence is why your group thinks the evidence matters. The justification of the evidence is important because scientists can use different kinds of evidence to support their claims. Your group will create your initial argument on a whiteboard. Your whiteboard should include all the information shown in Figure L17.3.

FIGURE L17.3 _____

Argument presentation on a whiteboard

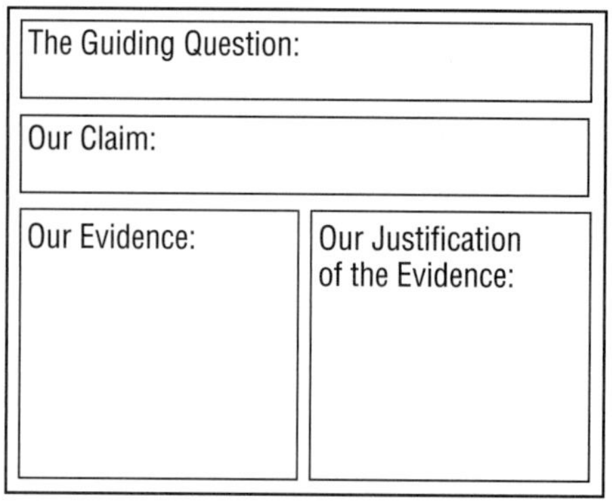

The Guiding Question:	
Our Claim:	
Our Evidence:	Our Justification of the Evidence:

Argumentation Session

The argumentation session allows all of the groups to share their arguments. One or two members of each group will stay at the lab station to share that group's argument, while the other members of the group go to the other lab stations to listen to and critique the other arguments. This is similar to what scientists do when they propose, support, evaluate, and refine new ideas during a poster session at a conference. If you are presenting your group's argument, your goal is to share your ideas and answer questions. You should also keep a record of the critiques and suggestions made by your classmates so you can use this feedback to make your initial argument stronger. You can keep track of specific critiques and suggestions for improvement that your classmates mention in the space below.

Critiques of our initial argument and suggestions for improvement:

If you are critiquing your classmates' arguments, your goal is to look for mistakes in their arguments and offer suggestions for improvement so these mistakes can be fixed. You should look for ways to make your initial argument stronger by looking for things that the other groups did well. You can keep track of interesting ideas that you see and hear during the argumentation in the space below. You can also use this space to keep track of any questions that you will need to discuss with your team.

Interesting ideas from other groups or questions to take back to my group:

Once the argumentation session is complete, you will have a chance to meet with your group and revise your initial argument. Your group might need to gather more data or design a way to test one or more alternative claims as part of this process. Remember, your goal at this stage of the investigation is to develop the best argument possible.

Report

Once you have completed your research, you will need to prepare an *investigation report* that consists of three sections. Each section should provide an answer for the following questions:

1. What question were you trying to answer and why?

2. What did you do to answer your question and why?

3. What is your argument?

Your report should answer these questions in two pages or less. You should write your report using a word processing application (such as Word, Pages, or Google Docs), if possible, to make it easier for you to edit and revise it later. You should embed any diagrams, figures, or tables into the document. Be sure to write in a persuasive style; you are trying to convince others that your claim is acceptable or valid.

References

National Aeronautics and Space Administration (NASA). n.d. Earth's energy budget. *https://earthobservatory.nasa.gov/Features/EnergyBalance/page4.php.*

National Aeronautics and Space Administration (NASA) Goddard Institute for Space Studies. 2016. GISS surface temperature analysis. *https://data.giss.nasa.gov/gistemp/graphs_v3.*

Factors That Affect Global Temperature
How Do Cloud Cover and Greenhouse Gas Concentration in the Atmosphere Affect
the Surface Temperature of Earth?

Checkout Questions

Lab 17. Factors That Affect Global Temperature: How Do Cloud Cover and Greenhouse Gas Concentration in the Atmosphere Affect the Surface Temperature of Earth?

1. Continued use of fossil fuels adds more greenhouse gases to the atmosphere each year. Does this have an impact on short-term temperature changes? In other words, will tomorrow be hotter than today because there will be more greenhouse gases in the atmosphere tomorrow?

2. Does human activity contribute to increased cloud cover? Explain your answer.

3. There are several different types of questions that scientists can ask as part of their investigations.

   a. I agree with this statement.
   b. I disagree with this statement.

   Explain your answer, using an example from your investigation about greenhouse gases and cloud cover.

LAB 17

4. Scientific knowledge can change over time.

   a. I agree with this statement.
   b. I disagree with this statement.

   Explain your answer, using an example from your investigation about greenhouse gases and cloud cover.

5. In science, it is critical to understand what makes a system stable or unstable and what controls rates of change in a system. Explain whether the system determining Earth's temperature is stable or unstable. Also, indicate what controls rates of change in Earth's temperature.

6. In science, it is important to track how energy moves into, within, and out of a system. Explain how energy moves into, out of, and within Earth's surface and atmosphere. It may help to draw a diagram. Also, indicate if there is a net gain or loss in energy for Earth's surface and for Earth's atmosphere.

Lab Handout

Lab 18. Carbon Dioxide Levels in the Atmosphere: How Has the Concentration of Atmospheric Carbon Dioxide Changed Over Time?

Introduction

There has been a lot of discussion about climate in recent years. This discussion usually focuses on average global temperature. In the United States some states have had above-average temperatures, some states have had below-average temperatures, and some states have had had near-average temperatures over the last 100 years. Figure L18.1 shows decadal temperature anomalies, or how the decadal average temperature for each state differs from the 20th-century average during three different decades. According to an ongoing temperature analysis conducted by scientists at the National Aeronautics and Space Administration (NASA) Goddard Institute for Space Studies, the average global temperature on Earth has increased by about 0.8°Celsius (1.4°Fahrenheit) since 1880 (see *https://data.giss.nasa.gov/gistemp/graphs_v3*).

A major contributing factor to global temperature is the concentration of carbon dioxide (CO_2) in the atmosphere. CO_2 is one type of greenhouse gas. Greenhouse gases trap heat from the Sun and warm the surface of Earth. Without greenhouse gases in the atmosphere, Earth would be too cold for humans to survive. As the concentration of greenhouse gases in the atmosphere increases, the temperature of Earth's surface will also increase.

Climate experts agree that human activity has significantly increased the amount of CO_2 in the atmosphere, leading to an overall rise in global temperatures. Some people, however,

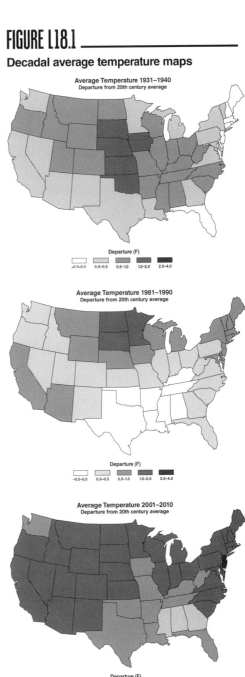

FIGURE L18.1

Decadal average temperature maps

Average Temperature 1931–1940
Departure from 20th century average

Departure (F)
-0.5–0.0   0.0–0.5   0.5–1.0   1.0–2.0   2.0–4.0

Average Temperature 1981–1990
Departure from 20th century average

Departure (F)
-0.5–0.0   0.0–0.5   0.5–1.0   1.0–2.0   2.0–4.0

Average Temperature 2001–2010
Departure from 20th century average

Departure (F)
-0.5–0.0   0.0–0.5   0.5–1.0   1.0–2.0   2.0–4.0

LAB 18

still question whether this increase is primarily due to human activity or to a natural process that causes climate change. There is research that shows global temperatures and atmospheric CO_2 levels have increased and decreased in a cyclical pattern for at least 650,000 years (Etheridge et al. 1998).

Before you can evaluate the merits of alternative explanations for the observed increase in average global temperature, it is important to understand how CO_2 levels have changed over Earth's history. You will therefore need to learn more about historical patterns of CO_2 levels.

Your Task

Analyze long-term historical data to determine whether CO_2 levels and average global temperature are changing at a different rate than they have in the past. Your goal is to use what you know about climate, patterns, and stability and change in systems to determine if human activity has made a significant change in global CO_2 levels and thus global temperature.

The guiding question of this investigation is: *How has the concentration of atmospheric carbon dioxide changed over time?*

Materials

You may use the following resources during your investigation:

- Average Global Temperature and Ice Core CO2 Data Excel file: This file provides information about changes in average global temperature over time and atmospheric CO_2 levels based on ice core samples that date back to 416,000 years before the present time.

- Climate Time Machine: This NASA website provides visualizations of current trends in sea ice levels, CO_2 levels, and global temperatures at *http://climate.nasa. gov/interactives/climate-time-machine*.

Safety Precautions

Be sure to follow all normal lab safety rules.

Investigation Proposal Required?   ☐ Yes   ☐ No

Getting Started

Scientists use some clever data sources to gain insight into Earth's history. One such data source is an ice core sample. To obtain an ice core sample, scientists drill down into a glacier or ice sheet and bring out a long cylindrical piece of ice (see Figure L18.2). Scientists can then count the layers in the ice core sample and determine how many years ago each

FIGURE L18.2 ⎯⎯⎯⎯⎯⎯⎯⎯⎯⎯⎯⎯

An example of an ice core

layer was on the surface of Earth. The ability to determine the age of layers by counting them is based on the law of superposition, which states that the oldest layers in a geologic sample are found at the bottom of the sample.

Scientists can also analyze the tiny air bubbles that are trapped in the ice at each layer of an ice core sample to determine the amounts of different gases that were in the atmosphere at the time that layer was created. When scientists make these measurements, they assume that natural laws operate today as they did in the past and that they will continue to do so in the future. Scientists therefore assume that the dissolved CO_2 levels in an ice core layer correspond to the amount of CO_2 present in the atmosphere at the time the ice was made, just like dissolved CO_2 levels in fresh ice match the CO_2 levels in the current atmosphere.

The Average Global Temperature and Ice Core CO2 Data Excel file includes information about the atmospheric CO_2 concentration and changes in average global temperature over time. The file includes two tabs:

1. The first tab, which is called "CO2 and Temp 1880-2016," includes atmospheric CO_2 levels and average global temperature anomalies from 1880 to 2016. The term *temperature anomaly* means the difference from the long-term average. A positive anomaly value indicates that the observed temperature was warmer than the long-term average, and a negative anomaly indicates that the observed temperature was cooler than the long-term average. Scientists calculate and report temperature anomalies because they more accurately describe climate variability than absolute temperatures do, and these anomalies make it easier to find patterns in temperature trends. The yearly temperature anomaly values come from the National Oceanic and Atmospheric Administration's National Centers for Environmental Information (see *www.ncdc.noaa.gov/monitoring-references/faq/anomalies.php*), the 1880–2004 atmospheric CO_2 levels come from Etheridge et al. (2010), and the 2005–2016 atmospheric CO_2 levels come from NASA's Global Climate Change website (see *https://climate.nasa.gov/vital-signs/carbon-dioxide*).

2. The second tab, which is called "CO2 Before 1880," includes atmospheric CO_2 levels dating back 416,000 years based on measurements taken from ice cores by Etheridge et al. (1998).

You can use the data from the Excel file to see how CO_2 concentrations and global temperature typically change over a very long time scale. As you analyze these data, think about the following questions:

- Will you need to analyze some data separately from others?
- What types of patterns might you look for as you analyze your data?
- What type of diagram could you create to help make sense of your data?
- How could you use mathematics to describe a change over time or if there is a relationship between variables?
- What type of graph could you create to help make sense of your data?

You can also use the visualizations on NASA's Climate Time Machine web page to examine how some of Earth's key climate indicators have changed in the recent past. This web page provides satellite pictures of the annual Arctic sea ice minimums dating back to 1979. At the end of each summer, the sea ice cover reaches its minimum extent, leaving what is called the perennial ice cover. The Climate Time Machine also shows global changes in the concentration and distribution of CO_2 in the atmosphere dating back to 2002 at an altitude range of 1.9–8 miles. The yellow-to-red regions indicate higher concentrations of CO_2, while the blue-to-green areas indicate lower concentrations, measured in parts per million. Finally, and perhaps most important, the Climate Time Machine provides a color-coded map that shows how global surface temperatures have changed dating back to 1884. Dark blue indicates areas cooler than average, and dark red indicates areas warmer than average.

Connections to the Nature of Scientific Knowledge and Scientific Inquiry

As you work through your investigation, be sure to think about

- how scientific knowledge can change over time, and
- the assumptions made by scientists about order and consistency in nature.

Initial Argument

Once your group has finished collecting and analyzing your data, your group will need to develop an initial argument. Your initial argument needs to include a claim, evidence to support your claim, and a justification of the evidence. The *claim* is your group's answer to the guiding question. The *evidence* is an analysis and interpretation of your data. Finally, the *justification* of the evidence is why your group thinks the evidence matters. The justification of the evidence is important because scientists can use different kinds of evidence to support their claims. Your group will create your initial argument on a whiteboard. Your whiteboard should include all the information shown in Figure L18.3.

FIGURE L18.3 _____

Argument presentation on a whiteboard

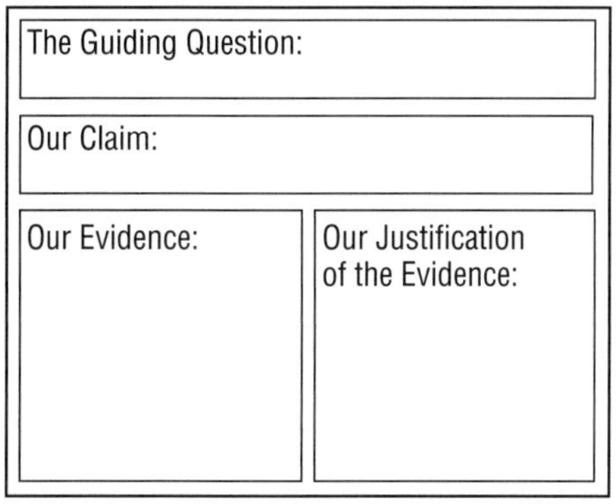

The Guiding Question:	
Our Claim:	
Our Evidence:	Our Justification of the Evidence:

Argumentation Session

The argumentation session allows all of the groups to share their arguments. One or two members of each group will stay at the lab station to share that group's argument, while the other members of the group go to the other lab stations to listen to and critique the other arguments. This is similar to what scientists do when they propose, support, evaluate, and refine new ideas during a poster session at a conference. If you are presenting your group's argument, your goal is to share your ideas and answer questions. You should also keep a record of the critiques and suggestions made by your classmates so you can use this feedback to make your initial argument stronger. You can keep track of specific critiques and suggestions for improvement that your classmates mention in the space below.

Critiques of our initial argument and suggestions for improvement:

If you are critiquing your classmates' arguments, your goal is to look for mistakes in their arguments and offer suggestions for improvement so these mistakes can be fixed. You should look for ways to make your initial argument stronger by looking for things that the other groups did well. You can keep track of interesting ideas that you see and hear during the argumentation in the space below. You can also use this space to keep track of any questions that you will need to discuss with your team.

LAB 18

Interesting ideas from other groups or questions to take back to my group:

Once the argumentation session is complete, you will have a chance to meet with your group and revise your initial argument. Your group might need to gather more data or design a way to test one or more alternative claims as part of this process. Remember, your goal at this stage of the investigation is to develop the best argument possible.

Report

Once you have completed your research, you will need to prepare an *investigation report* that consists of three sections. Each section should provide an answer for the following questions:

1. What question were you trying to answer and why?

2. What did you do to answer your question and why?

3. What is your argument?

Your report should answer these questions in two pages or less. You should write your report using a word processing application (such as Word, Pages, or Google Docs), if possible, to make it easier for you to edit and revise it later. You should embed any diagrams, figures, or tables into the document. Be sure to write in a persuasive style; you are trying to convince others that your claim is acceptable or valid.

References

Etheridge, D. M, L. P. Steele, R. L. Langenfelds, R. J. Francey, J. Barnola, V. I. and Morgan. 1998. Historical CO_2 records from the Law Dome DE08, DE08-2, and DSS ice cores. In *Trends: A compendium of data on global change*. Oak Ridge, TN: U.S. Department of Energy, Oak Ridge National Laboratory, Carbon Dioxide Information Analysis Center. Available at *www.co2.earth/co2-ice-core-data*.

Etheridge, et al. 2010. Law Dome Ice Core 2000-Year CO2, CH4, and N2O Data. IGBP PAGES/World Data Center for Paleoclimatology Data Contribution Series 2010-070. Boulder, CO: NOAA/NCDC Paleoclimatology Program. Available at *ftp://ftp.ncdc.noaa.gov/pub/data/paleo/icecore/antarctica/law/law2006.txt*.

Checkout Questions

Lab 18. Carbon Dioxide Levels in the Atmosphere: How Has the Concentration of Atmospheric Carbon Dioxide Changed Over Time?

1. Sketch a graph of the how the concentration of atmospheric carbon dioxide has changed over time.

2. A scientist collects yearly global average temperature and compiles it into the table below.

Year	Average temperature
1880	56.8°F
1900	56.9°F
1920	56.7°F
1930	56.9°F
1940	57.3°F
1960	57.1°F
1980	57.6°F
2000	57.9°F
2010	58.4°F

   a. What is the rate of change for the time period between1880 and 1930 and the time period between 1960 and 2010?

   b. Are the rates of change significantly different from one another?

c. How do you know?

d. What additional information would you need to determine whether the global climate is stable?

3. Once new scientific knowledge is developed, it will not be abandoned or modified in light of new evidence.

   a. I agree with this statement.

   b. I disagree with this statement.

   Explain your answer, using an example from your investigation about carbon dioxide levels in the atmosphere.

4. Science assumes that objects and events in natural systems occur in consistent patterns that are understandable through measurement and observation.

   a. I agree with this statement.
   b. I disagree with this statement.

   Explain your answer, using an example from your investigation about carbon dioxide levels in the atmosphere.

5. It is critical for scientists to be able to recognize what is relevant at different time frames and scales. Explain why analyzing data in the context of appropriate time frames and scales is important, using an example from your investigation about carbon dioxide levels in the atmosphere.

6. It is critical to understand what makes a system stable or unstable and what controls rates of change in a system. Explain why it is important to determine whether a system is changing or is stable, using an example from your investigation about trends in average global temperatures.

Application Lab

Lab Handout

Lab 19. Differences in Regional Climate: Why Do Two Cities Located at the Same Latitude and Near a Body of Water Have Such Different Climates?

Introduction

Weather describes the current atmospheric conditions at a particular location. *Climate,* in contrast, is the aggregate or typical weather for a particular location over a long period of time. People often describe the climate of a region by reporting average temperatures and rainfall by month or by season. Cities located at higher latitudes (i.e., farther from the equator) experience greater changes in day length and Sun angle over the course of a year. These cities, as a result, have greater seasonal temperature differences than cities that are located closer to the equator. It is therefore not surprising that cities located at different latitudes often have very different climates. Cities located at the same latitude, however, can also have very different climates. A good example of this phenomenon is seen when we look at the climates of San Francisco, California, and Norfolk, Virginia (see Figure L19.1).

FIGURE L19.1

Locations of San Francisco and Norfolk

San Francisco is located on the coast of the Pacific Ocean at 37.7° N latitude. It has mild summers and winters; the average high temperature between 1945 and 2017 has been 17.1°C (62.8°F) for July and 9.7°C (49.4°F) for January. San Francisco is very dry, averaging 19.7 inches of rain per year from 1946 to 2017. (Climate information from the National Oceanic and Atmospheric Administration [NOAA] National Centers for Environmental Information is available at *www.ncdc.noaa.gov.*) According to the Köppen Climate Classification System (the most widely used system for classifying the world's climates; see *www.britannica.com/science/Koppen-climate-classification*), San Francisco has a temperate Mediterranean climate with warm summers (denoted Csb in the Köppen system). In contrast, Norfolk, which is located on the Atlantic Ocean at 36.9° N latitude, has hot summers and mild winters. The average high temperature between 1945 and 2017 has been 30.9°C (87.6°F) in July and 9.5°C (49.1°F) in January. Between 1945 and 2017, Norfolk has averaged 46.4 inches of rain a year. Norfolk has a humid subtropical climate (Cfa) based on the Köppen classification system. In summary, San Francisco and Norfolk have very different temperature and precipitation patterns throughout the year even though they are located at similar latitudes. To understand why these two cities have such different temperature and precipitation patterns, we must consider all the different factors than can affect the climate of a region.

There are at least six important factors to consider when someone attempts to explain a difference in two or more regional climates:

- *Latitude,* as noted earlier, determines changes in day length and sun angle throughout the year.

- *Elevation,* which is the height of an area above sea level. Generally, as elevation increases, temperature decreases.

- *Proximity to a large body of water.* Land heats up and cools down faster than water. Water can also store more heat energy than land. This makes the climate of a region that is located near a large body of water, such as an ocean, more moderate because the water absorbs extra heat energy during the summer and releases heat into the air during winter.

- *The nature of nearby ocean currents.* Ocean currents move large amounts of water with different properties to different locations across the Earth. Winds, tides, and differences in water temperature and salinity at different locations in the ocean affect the path an ocean current follows over time.

- *The direction and strength of prevailing winds.* Winds can move air masses with specific properties from a source region to a different region. In the Northern Hemisphere, winds tend to blow from west to east (westerly winds) in the midlatitudes and northeast to southwest between the Tropic of Cancer and the equator.

- *Local topography.* The presence of absence of a mountain in a region, for example, can affect precipitation patterns and therefore climate.

These six factors can help us understand why there are different climates at different locations around the globe. Yet, some of them may or may not be useful when we need to explain the different temperature and precipitation patterns observed in cities such as San Francisco and Norfolk. The two cities, as noted earlier, have much in common. San Francisco and Norfolk are located at similar latitudes, at an elevation slightly above sea level, and are near an ocean but on different coasts. You will therefore need to learn more about the nature of nearby ocean currents, the nature of any prevailing winds in these regions, and the local topography around these cities to figure out why these two cities have such different climates. Next, you will put all these pieces of information together to develop a conceptual model that not only explains the different climates in San Francisco and Norfolk but can also explain differences in the climates of other cities that are located at similar latitudes.

Your Task

Develop a conceptual model that you can use to explain the temperature and precipitation patterns in San Francisco and Norfolk. Your conceptual model must reflect what we know about the various factors that can affect climate, patterns, and systems and system models. To be considered valid or acceptable, you should be able use your conceptual model to not only explain why San Francisco and Norfolk have different climates but also to predict the temperature and precipitation patterns of several other pairs of cities that are located at similar latitudes on Earth.

The guiding question of this investigation is, **Why do two cities located at the same latitude and near a body of water have such different climates?**

Materials

You can use the following online resources during your investigation:

- U.S. Climate Data. You can access this website, which includes detailed climate data and the location of most major U.S. Cities, at *www.usclimatedata.com/climate/united-states/us.*

- Wind rose data from the National Oceanic and Atmospheric Administration (NOAA). You can access this database, which includes monthly wind speed and direction information for 237 U.S. cities, at *www.climate.gov/maps-data/dataset/monthly-wind-rose-plots-charts.*

- *Earth: A Global Map of Wind, Weather, and Ocean Conditions.* You can access this interactive animated map that shows current wind speeds and direction for the entire planet at *https://earth.nullschool.net.*

- *My NASA Data.* You can access this database, which includes information about ocean surface temperatures and the average wind speed and direction by month over the entire year, at *https://mynasadata.larc.nasa.gov.*

- *State of the Ocean* (SOTO). You can access this visualization tool, which includes information about current and past ocean currents and changes in surface temperature, through the NASA Jet Propulsion Laboratory Physical Oceanography Distributed Active Archive Center at *https://podaac.jpl.nasa.gov*. Click on "Data Access" and "SOTO (State of the Ocean)" to open the visualization tool.

Safety Precautions

Follow all normal lab safety rules.

Investigation Proposal Required? ☐ Yes ☐ No

Getting Started

The first step in developing a conceptual model that explains differences in the temperature and precipitation patterns of San Francisco and Norfolk is to collect information about seasonal changes in temperature and precipitation in both cities. This information can be found at the U.S. Climate Data website. Be sure look for any patterns that you can use to help develop your conceptual model. Next, you can learn about the prevailing winds at each location using data provided by the NOAA wind rose website and the Earth: A Global Map of Wind, Weather, and Ocean Conditions website (which provides real-time data you can use). You can also access data about ocean temperatures and wind patterns by using the My NASA data. The final website, SOTO, will allow you to visualize a variety of ocean characteristics on a map of the world.

To learn more about how prevailing winds affect climate, you must first determine what type of data you need to collect, how you will collect it, and how you will analyze it. To determine *what type of data you need to collect,* think about the following questions:

- What are the boundaries and components of the system you are studying?
- How do the components of the system interact with each other?
- When is this system stable, and under which conditions does it change?
- What could be the underlying cause of this phenomenon?
- What type of measurements or observations will you need?
- What types of patterns could you look for in the available data?

To determine *how you will collect your data,* think about the following questions:

- What conditions need to be satisfied to establish a cause-and-effect relationship?
- How can you describe the components of the system quantitatively?
- What measurement scale or scales should you use to collect data?
- What type of comparisons will you need to make?
- How will you keep track of and organize the data you collect?

To determine *how you will analyze your data,* think about the following questions:

- What types of patterns might you look for as you analyze your data?
- How could you use mathematics to document a difference between conditions?
- What type of comparisons and calculations will you need to make?
- What type of graph could you create to help make sense of your data?

Once you feel you have gathered sufficient data and identified important patterns about how oceans and prevailing winds affect climate, your group can develop a conceptual model that can be used to explain why San Francisco and Norfolk have such different climates. To be valid or acceptable, your conceptual model must be able to explain

    a. why San Francisco and Norfolk have such different seasonal temperatures, and

    b. why San Francisco and Norfolk have such different rain patterns.

The last step in your investigation will be to generate the evidence that you need to convince others that your conceptual model is valid or acceptable. To accomplish this goal, you will use your model to predict the temperature and precipitation patterns in several additional cities. These cities should be ones that you have not looked up before but are located at similar latitudes on different coasts. Some good pairs of cities to compare are

- San Diego, California, and Charleston, South Carolina;
- Portland, Oregon, and Bangor, Maine;
- Eureka, California, and New York City; and
- Santa Monica, California, and Wilmington, North Carolina.

You can also attempt to show how using a different version of your model or making a specific change to a portion of your model will make your model inconsistent with data you have or the facts we know about climate. Scientists often make comparisons between different versions of a model in this manner to show that a model is valid or acceptable. If you are able to use your conceptual model to make accurate predictions about the climates of other cities or if you are able show how your conceptual model explains the climates of different cities better than other conceptual models, then you should be able to convince others that it is valid or acceptable.

Connections to the Nature of Scientific Knowledge and Scientific Inquiry

As you work through your investigation, be sure to think about

- how models are used as tools for reasoning about phenomena, and
- how the culture of science, societal needs, and current events influence the work of scientists.

Initial Argument

Once your group has finished collecting and analyzing your data, your group will need to develop an initial argument. Your initial argument needs to include a claim, evidence to support your claim, and a justification of the evidence. The *claim* is your group's answer to the guiding question. The *evidence* is an analysis and interpretation of your data. Finally, the *justification* of the evidence is why your group thinks the evidence matters. The justification of the evidence is important because scientists can use different kinds of evidence to support their claims. Your group will create your initial argument on a whiteboard. Your whiteboard should include all the information shown in Figure L19.2.

Argumentation Session

The argumentation session allows all of the groups to share their arguments. One or two members of each group will stay at the lab station to share that group's argument, while the other members of the group go to the other lab stations to listen to and critique the other arguments. This is similar to what scientists do when they propose, support, evaluate, and refine new ideas during a poster session at a conference. If you are presenting your group's argument, your goal is to share your ideas and answer questions. You should also keep a record of the critiques and suggestions made by your classmates so you can use this feedback to make your initial argument stronger. You can keep track of specific critiques and suggestions for improvement that your classmates mention in the space below.

Critiques of our initial argument and suggestions for improvement:

FIGURE L19.2

Argument presentation on a whiteboard

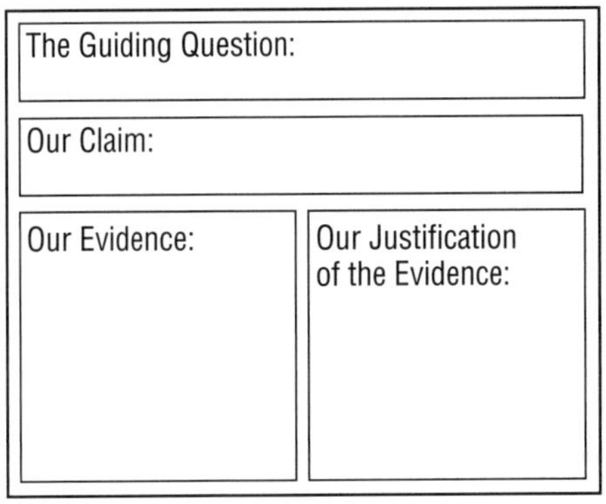

The Guiding Question:	
Our Claim:	
Our Evidence:	Our Justification of the Evidence:

If you are critiquing your classmates' arguments, your goal is to look for mistakes in their arguments and offer suggestions for improvement so these mistakes can be fixed. You should look for ways to make your initial argument stronger by looking for things that the other groups did well. You can keep track of interesting ideas that you see and hear during the argumentation in the space below. You can also use this space to keep track of any questions that you will need to discuss with your team.

Interesting ideas from other groups or questions to take back to my group:

Once the argumentation session is complete, you will have a chance to meet with your group and revise your initial argument. Your group might need to gather more data or design a way to test one or more alternative claims as part of this process. Remember, your goal at this stage of the investigation is to develop the best argument possible.

Report

Once you have completed your research, you will need to prepare an investigation report that consists of three sections. Each section should provide an answer for the following questions:

1. What question were you trying to answer and why?

2. What did you do to answer your question and why?

3. What is your argument?

Your report should answer these questions in two pages or less. You should write your report using a word processing application (such as Word, Pages, or Google Docs), if possible, to make it easier for you to edit and revise it later. You should embed any diagrams, figures, or tables into the document. Be sure to write in a persuasive style; you are trying to convince others that your claim is acceptable or valid.

Checkout Questions

Lab 19. Differences in Regional Climate: Why Do Two Cities Located at the Same Latitude and Near a Body of Water Have Such Different Climates?

1. What is climate?

2. How do prevailing winds and ocean currents affect the climates of different regions?

3. The map below shows the locations of three cities: Los Angeles, California; Oklahoma City, Oklahoma; and Myrtle Beach, South Carolina. The map also includes information about prevailing wind directions (dotted lines) and the location of cold water (solid lines) and warm water (dashed lines) ocean currents.

LAB 19

a. Rank the cities based on their summer high temperatures, with 1 being the coolest summer, and 3 being the warmest summer.

City	Rank
Los Angeles, CA	_____
Oklahoma City, OK	_____
Myrtle Beach, SC	_____

b. How do you know? Explain why you ranked the cities this way.

   c. Rank the cities based on their winter low temperatures, with 1 being the warmest winter, and 3 being the coldest winter.

City	Rank
Los Angeles, CA	_____
Oklahoma City, OK	_____
Myrtle Beach, SC	_____

   d. How do you know? Explain why you ranked the cities this way.

4. Scientists conduct investigation based on their own interests; societal needs and current events do not influence the work of scientists at all.

   a. I agree with this statement.

   b. I disagree with this statement.

Explain your answer, using an example from your investigation about differences in regional climate.

5. Scientists use models as tools for reasoning about natural phenomena.

    a. I agree with this statement.
    b. I disagree with this statement.

    Explain your answer, using an example from your investigation about differences in regional climate.

6. Scientists often need to look for patterns that occur in the data they collect and analyze. Explain why identifying patterns are important, using an example from your investigation about differences in regional climate.

7. Defining a system under study and making a model of it are tools for developing a better understanding of natural phenomena in science. Give an example of how system models are used by scientists to investigate natural and designed systems.

SECTION 6
Human Impact

Introduction Labs

Lab 20. Predicting Hurricane Strength: How Can Someone Predict Changes in Hurricane Wind Speed Over Time?

Introduction

The strongest tropical storms are called hurricanes, typhoons, or cyclones. The different names all mean the same thing but are used to describe tropical storms that originate in different parts of the world. If a huge storm starts off the west coast of Africa in the Atlantic, it is called a hurricane. Hurricanes have strong winds, a spiral shape, and a low-pressure center called an eye. Unlike other natural hazards, such as earthquakes or even tornadoes, we can observe the development of a hurricane over time and track how it moves across the ocean.

A hurricane begins as a tropical disturbance in the ocean off the west coast of Africa. A tropical disturbance forms in an area where the ocean surface temperature is at least 27°C (80°F). The warm humid air at that location rises and creates an area of low atmospheric pressure near the ocean surface. Cooler air in the region then rushes into the area of low pressure. This air picks up evaporated water from the surface, increases in temperature, and moves upward into the atmosphere. This process produces large water-filled thunderclouds around the area of low pressure. The trade winds (which blow from east to west) slowly push the disturbance to the west.

Over the next few days, more warm air will rise and the winds will begin to circulate around the center of the disturbance in counterclockwise (when viewed from above) direction. A layer of clouds called an outflow will also begin to form at the top of the storm. Winds inside the storm will increase in speed over time. When the winds within the storm are between 25 and 38 miles per hour (mph), the storm is called a tropical depression. When the wind speeds reach 39 mph, the storm is classified as a tropical storm rather than as a tropical depression. This is also the point in time when the storm gets a name. In a couple of days, as the system moves across the ocean, the clouds expand and the winds continue to speed up. When the wind speeds inside the storm reach 74 mph, it is classified as a hurricane (see Figure L20.1).

FIGURE L20.1

Image of Hurricane Isabel about 400 miles north of Puerto Rico on September 14, 2003, captured by the NASA Terra satellite; the sustained wind speed inside Hurricane Isabel at that time was 155 mph

Note: A full-color version of this figure is available on the book's Extras page at *www.nsta.org/adi-ess*.

In an average year, several different hurricanes will form over the Atlantic Ocean and head westward toward the Caribbean, the east coast of Central America, or the southeastern United States. Figure L20.2 shows the tracks of all North Atlantic Ocean hurricanes that developed between 1980 and 2005. The points on each track represent the location of that storm at six-hour intervals. Hurricanes will often last several weeks before they break down because they tend to move very slowly across the ocean. In fact, hurricanes usually travel across the ocean at only about 24 kilometers per hour (or 15 mph).

FIGURE L20.2

Map showing the tracks of all hurricanes in the North Atlantic Ocean from 1980 to 2005; the points show the locations of the storms at six-hour intervals

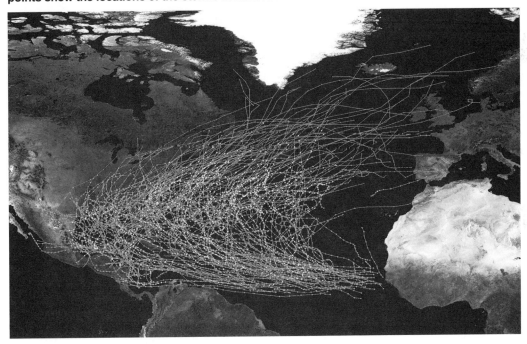

Note: A full-color version of this figure is available on the book's Extras page at *www.nsta.org/adi-ess.*

Scientists use the sustained wind speed inside a hurricane to classify it. The sustained wind speed inside a hurricane, however, can increase or decrease over time. It is therefore important for scientists to understand why the winds inside a hurricane change over time. This type of information is important because scientists are responsible for issuing evacuation warnings, and they need to know if the wind within a hurricane is likely to increase or decrease before it reaches landfall. It is also important to understand the factors that affect the wind speed of a hurricane over time as it moves over water or land; this information helps city planners to establish building codes for cities to ensure that new buildings will be able to withstand the winds of a typical hurricane for that area.

LAB 20

In this investigation, you will have an opportunity to learn more about the factors that affect wind speed within hurricanes. Your goal is to develop a conceptual model that you can use to not only explain why the wind speed within a hurricane changes over time as it moves over water or land but also predict how the strength of a hurricane will increase or decrease over time based on the path that it follows.

Your Task

Develop a conceptual model that can be used to explain why wind speed inside a hurricane changes over time as it moves over water or land. Your conceptual model must be based on what we know about natural hazards; weather; the importance of tracking how energy flows into, within, and out of a system; and cause-and-effect relationships. Once you have developed your model, you will need to test it to see if you can use it to make accurate predictions about how the strength of several hurricanes changed over time in the past.

The guiding question of this investigation is, *How can someone predict changes in hurricane wind speed over time?*

Materials

You may use any of the following materials during your investigation:

Equipment
- Computer or tablet with internet access

Other Resources
- Some Major Hurricanes handout
- Hurricane Track A—Black and White handout (use to test your model)
- Hurricane Track A—Color handout (use to check your predictions)
- Hurricane Track B—Black and White handout (use to test your model)
- Hurricane Track B—Color handout (use to check your predictions)

Safety Precautions

Follow all normal lab safety rules.

Investigation Proposal Required?   ☐ Yes   ☐ No

Getting Started

The first step in this investigation is to determine how the strength of a hurricane changes over time as it travels over water and land. To accomplish this goal, you will need to examine several different historical hurricane tracks and look for patterns that you can use to explain and predict changes in wind speed. You can access historical hurricane track maps from the National Oceanic and Atmospheric Administration (NOAA) at *https://coast.noaa.gov/hurricanes*. The tracks on the maps are color-coded using the Saffir-Simpson Hurricane Scale (see Table L20.1) so you can keep track of how the strength of a storm changed over time.

TABLE L20.1

Types of tropical storms classified by wind speed according to the Saffir-Simpson Hurricane Scale and by colors used on historical hurricane track maps from the National Oceanic and Atmospheric Administration

Classification	Wind speed			Color used on historical hurricane track maps
	mph	kn	km/h	
Category 5 hurricane	≥ 157	≥ 137	≥ 252	Purple
Category 4 hurricane	130–156	113–136	209–251	Pink
Category 3 hurricane	111–129	96–112	178–208	Red
Category 2 hurricane	96–110	83–95	154–177	Orange
Category 1 hurricane	74–95	64–-82	119–153	Yellow
Tropical storm	39–73	34–63	63–118	Green
Tropical depression	25–38	22–33	40–62	Blue

You can then examine how hurricane wind speed is related to the surface temperature of the ocean (also called sea surface temperature, or SST). Information about current SSTs in the North Atlantic can be found at *www.ospo.noaa.gov/Products/ocean/sst/contour/index.html*. You can also find information about monthly SSTs for 1984–1998 at *www.ospo.noaa.gov/Products/ocean/sst/monthly_mean.html*. Finally, you may want to compare land surface temperature to SSTs. Information about land surface temperature during the daytime by month can be found at *http://earthobservatory.nasa.gov/GlobalMaps*.

Once you finished analyzing these data, you can develop your conceptual model. A conceptual model is an idea or set of ideas that explains what causes a particular phenomenon in nature. People often use words, images, and arrows to describe a conceptual model. Your conceptual model needs to be able to explain why hurricanes wind speed changes over time. The model also needs to be consistent with what we know about natural hazards, weather, and how energy flows into, within, and out of systems.

The last step in this investigation is to test your model. To accomplish this goal, you can use your model to make predictions about how the wind speed of past hurricanes changed over time using the Hurricane Track A and Hurricane Track B handouts. The black-and-white track maps include letters that mark specific locations along these tracks. You goal is to predict the category of these hurricanes at these locations. Your teacher will then give you color versions of these hurricane track maps. The color versions include information about the strength of these hurricanes at each location. You can use these maps to determine if your predictions were accurate. If you are able to make accurate predictions about how the wind speed within these two hurricanes changed over time, then you will be able to generate the evidence you need to convince others that the conceptual model you developed is valid or acceptable.

Connections to the Nature of Scientific Knowledge and Scientific Inquiry

As you work through your investigation, be sure to think about

- the use of models as tools for reasoning about natural phenomena, and
- the types of questions that scientists can investigate.

Initial Argument

Once your group has finished collecting and analyzing your data, your group will need to develop an initial argument. Your initial argument needs to include a claim, evidence to support your claim, and a justification of the evidence. The *claim* is your group's answer to the guiding question. The *evidence* is an analysis and interpretation of your data. Finally, the *justification* of the evidence is why your group thinks the evidence matters. The justification of the evidence is important because scientists can use different kinds of evidence to support their claims. Your group will create your initial argument on a whiteboard. Your whiteboard should include all the information shown in Figure L20.3.

FIGURE L20.3

Argument presentation on a whiteboard

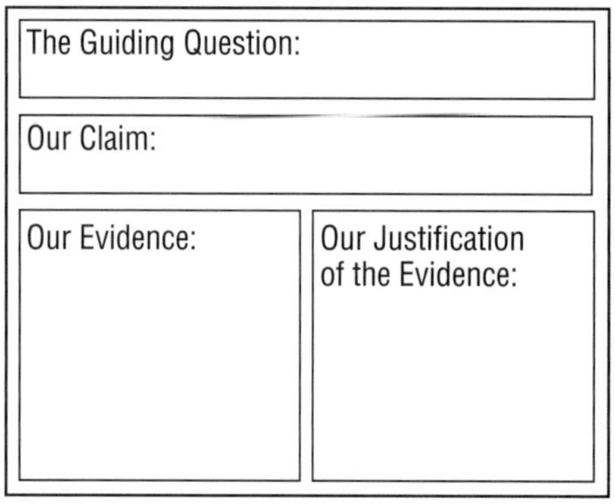

The Guiding Question:	
Our Claim:	
Our Evidence:	Our Justification of the Evidence:

Argumentation Session

The argumentation session allows all of the groups to share their arguments. One or two members of each group will stay at the lab station to share that group's argument, while the other members of the group go to the other lab stations to listen to and critique the other arguments. This is similar to what scientists do when they propose, support, evaluate, and refine new ideas during a poster session at a conference. If you are presenting your group's argument, your goal is to share your ideas and answer questions. You should also keep a record of the critiques and suggestions made by your classmates so you can use this feedback to make your initial argument stronger. You can keep track of specific critiques and suggestions for improvement that your classmates mention in the space below.

Critiques of our initial argument and suggestions for improvement:

If you are critiquing your classmates' arguments, your goal is to look for mistakes in their arguments and offer suggestions for improvement so these mistakes can be fixed. You should look for ways to make your initial argument stronger by looking for things that the other groups did well. You can keep track of interesting ideas that you see and hear during the argumentation in the space below. You can also use this space to keep track of any questions that you will need to discuss with your team.

Interesting ideas from other groups or questions to take back to my group:

Once the argumentation session is complete, you will have a chance to meet with your group and revise your initial argument. Your group might need to gather more data or design a way to test one or more alternative claims as part of this process. Remember, your goal at this stage of the investigation is to develop the best argument possible.

Report

Once you have completed your research, you will need to prepare an investigation report that consists of three sections. Each section should provide an answer for the following questions:

1. What question were you trying to answer and why?

2. What did you do to answer your question and why?

3. What is your argument?

Your report should answer these questions in two pages or less. You should write your report using a word processing application (such as Word, Pages, or Google Docs), if possible, to make it easier for you to edit and revise it later. You should embed any diagrams, figures, or tables into the document. Be sure to write in a persuasive style; you are trying to convince others that your claim is acceptable or valid.

LAB 20

Checkout Questions

Lab 20. Predicting Hurricane Strength: How Can Someone Predict Changes in Hurricane Wind Speed Over Time?

1. The maps below show the path of two different hurricanes. David was classified as a category 2 hurricane when it reached Florida. Andrew, in contrast, was classified as a category 4 hurricane when it passed over Florida 13 years later. Both hurricanes began as a tropical depression near Africa.

Path of Hurricane David (1979)   Path of Hurricane Andrew (1992)

   a. What are two factors that can affect the wind speed of a hurricane?

   b. Why was the sustained wind speed of Hurricane David less than then sustained wind speed of Hurricane Andrew when these two hurricanes made landfall in Florida?

National Science Teachers Association

2. Scientists create pictures of things to teach people about them. These pictures are models.

    a. I agree with this statement.

    b. I disagree with this statement.

Explain your answer, using an example from your investigation about hurricanes.

3. All questions can be answered by science.

    a. I agree with this statement.

    b. I disagree with this statement.

Explain your answer, using an example from your investigation about hurricanes.

4. Natural phenomena have causes, and uncovering causal relationships is a major activity of science. Explain why identifying cause-and-effect relationships is important, using an example from your investigation about hurricanes.

5. Tracking energy as it moves into, out of, and within systems is an important activity in science. Give an example for energy moving into, within, or out of a system from your investigation about hurricanes.

Lab Handout

Lab 21. Forecasting Extreme Weather: When and Under What Atmospheric Conditions Are Tornadoes Likely to Develop in the Oklahoma City Area?

Introduction

A tornado is a violent rotating column of air that extends from the clouds to the ground. The wind speeds in a tornado can reach as high as 480 kilometers per hour (almost 300 miles per hour). A tornado can destroy large buildings, uproot trees, and hurl vehicles hundreds of yards. The Oklahoma City metropolitan area is one spot in the United States that is known for frequent tornadoes. According to the National Weather Service, at least 162 tornadoes have touched down in the Oklahoma City metropolitan area between 1890 and 2013, an average of just over one per year (National Weather Service 2017). An example of a tornado in this area was the one that hit Moore, Oklahoma, on May 20th, 2013 at 2:46 p.m. (CDT) (National Weather Service n.d.; Thompson 2013). It stayed on the ground for 47 minutes, traveled 22.5 km (14 miles), and was 1.7 km (1.1 miles) wide at its peak (see Figure L21.1). The 2013 Moore tornado killed 24 people, injured another 212, destroyed 1,150 homes, and caused $2 billion in damage (see Figure L21.2).

FIGURE L21.1 _____

The 2013 EF5 Moore tornado as it passed through south Oklahoma City

FIGURE L21.2 _____

An overhead view of damage done by the 2013 Moore tornado

Tornadoes tend to develop within a supercell, which is a thunderstorm with a large rotating updraft. Thunderstorms tend to develop at locations where there is an interaction between a warm air mass and a cold air mass. Two air masses, however, can interact with each other in many different ways. For example, a warm air mass can move into an area that was formerly covered by cold air mass. A cold air mass can also move into an area that was occupied by a warm air mass. A warm air mass and a cold air mass can also move parallel to each other. These are just a few examples of how warm and cold air mass can interact with each other. Some of these interactions might lead to the development of a thunderstorm and some might not. Tornadoes also tend to develop in different regions during certain times of the year and at certain times of the day. People can make better forecasts about when tornadoes are likely to develop if they understand when tornadoes tend to happen and the atmospheric conditions that tend to be associated with them.

There are still many questions to be answered about what causes a tornado to develop and the factors that affect the wind speed of a tornado. One of the best ways to learn more about tornadoes is to keep a record of which months they happen, what time of day they happen, and what the atmospheric conditions are like in a region right before they develop and then look for patterns that might give us clues about when they are likely to develop and why. This information can then be used to help forecast future tornadoes and to inform the development of new technologies that can mitigate the damage that often results from this type of catastrophic event. In this investigation, you will have an opportunity to use historical weather records to determine when tornadoes tend to occur in the Oklahoma City area and what the atmospheric conditions were like in the region at that time. You can then use this information to identify a way to help predict when a tornado will likely appear in or around Oklahoma City.

Your Task

Use what you know about natural hazards, weather, patterns, and cause-and-effect relationships to analyze historical weather data from the Oklahoma City area to determine when tornadoes tend to happen in this area and what the atmospheric conditions tend to be like at that time. Your goal is to help people in the Oklahoma City area make better forecasts about these potentially catastrophic events.

The guiding question of this investigation is, *When and under what atmospheric conditions are tornadoes likely to develop in the Oklahoma City area?*

Materials

You can use the following online resources during your investigation:

- The National Oceanic and Atmospheric Administration (NOAA) National Weather Service provides information about every recorded tornado in the Oklahoma City area at *www.weather.gov/oun/tornadodata*.

- NOAA Central Library's Daily Weather Map Archive provides historical U.S. weather maps at *www.wpc.ncep.noaa.gov/dwm/dwm.shtml*.

- Weather Underground provides historical weather data at *www.wunderground.com/history*.

Safety Precautions

Follow all normal lab safety rules.

Investigation Proposal Required?   ☐ Yes   ☐ No

Getting Started

To answer the guiding question, you will need to analyze historical weather data. The first step in your analysis is to learn more about the tornadoes that have hit the Oklahoma City area. You will need to know when during the year these tornadoes developed, what time of the day they happened, how long they lasted, the path they followed, and their magnitude. NOAA provides this information about every recorded tornado in the Oklahoma City area since 1890. You can decide which ones and how many to study.

The next step in your investigation will be to learn more about what the atmospheric conditions were like in the Oklahoma City area before these different tornadoes developed. To accomplish this goal, you can use the NOAA Central Library's Daily Weather Map Archive or Weather Underground. The NOAA Central Library's Daily Weather Map Archive allows you to look up and download the weather maps for the entire United States for a specific date in the past. The Weather Underground website allows you to look up detailed information about the weather in any city dating back to 1945. You can use these two websites to investigate the changes in weather conditions near the dates of the tornadoes listed in the Oklahoma City Area Tornado Table. When you access these websites, you need to think about what data you need to collect, how you will collect it, and how you will analyze it.

To determine *what type of data you need to collect*, think about the following questions:

- Of all of the information you can access, which data are relevant and which data are irrelevant?
- How will you decide which tornadoes to include and which ones to exclude in your study?

To determine *how you will collect the data*, think about the following questions:

- How much data do you need to sufficiently answer your question?
- What scale or scales should you use?
- How will you keep track of and organize the data you collect?
- How will you organize your data?

To determine *how you will analyze the data*, think about the following questions:

- What types of patterns could you look for in your data?
- How could you use mathematics to describe a change over time?

LAB 21

- What type of table or chart could you create to help make sense of your data?
- How could you use mathematics to describe a relationship between variables?

Connections to the Nature of Scientific Knowledge and Scientific Inquiry

As you work through your investigation, be sure to think about

- the difference between data and evidence in science, and
- how scientists use different methods to answer different types of questions.

Initial Argument

Once your group has finished collecting and analyzing your data, your group will need to develop an initial argument. Your initial argument needs to include a claim, evidence to support your claim, and a justification of the evidence. The *claim* is your group's answer to the guiding question. The *evidence* is an analysis and interpretation of your data. Finally, the *justification* of the evidence is why your group thinks the evidence matters. The justification of the evidence is important because scientists can use different kinds of evidence to support their claims. Your group will create your initial argument on a whiteboard. Your whiteboard should include all the information shown in Figure L21.3.

FIGURE L21.3 _____

Argument presentation on a whiteboard

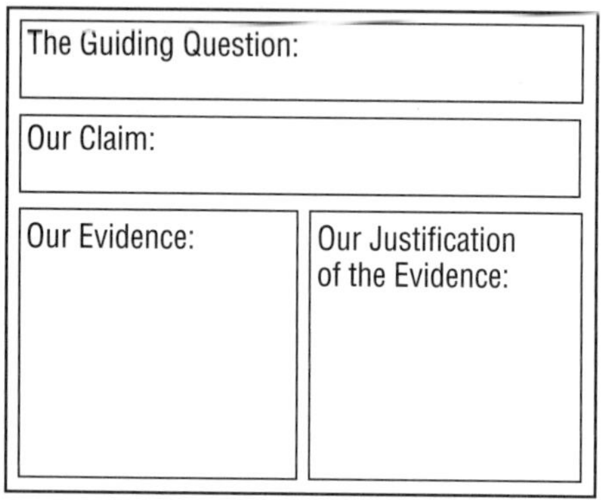

The Guiding Question:	
Our Claim:	
Our Evidence:	Our Justification of the Evidence:

Argumentation Session

The argumentation session allows all of the groups to share their arguments. One or two members of each group will stay at the lab station to share that group's argument, while the other members of the group go to the other lab stations to listen to and critique the other arguments. This is similar to what scientists do when they propose, support, evaluate, and refine new ideas during a poster session at a conference. If you are presenting your group's argument, your goal is to share your ideas and answer questions. You should also keep a record of the critiques and suggestions made by your classmates so you can use this feedback to make your initial argument stronger. You can keep track of specific critiques and suggestions for improvement that your classmates mention in the space provided.

Critiques of our initial argument and suggestions for improvement:

If you are critiquing your classmates' arguments, your goal is to look for mistakes in their arguments and offer suggestions for improvement so these mistakes can be fixed. You should look for ways to make your initial argument stronger by looking for things that the other groups did well. You can keep track of interesting ideas that you see and hear during the argumentation in the space below. You can also use this space to keep track of any questions that you will need to discuss with your team.

Interesting ideas from other groups or questions to take back to my group:

Once the argumentation session is complete, you will have a chance to meet with your group and revise your initial argument. Your group might need to gather more data or design a way to test one or more alternative claims as part of this process. Remember, your goal at this stage of the investigation is to develop the best argument possible.

Report

Once you have completed your research, you will need to prepare an *investigation report* that consists of three sections. Each section should provide an answer for the following questions:

1. What question were you trying to answer and why?

2. What did you do to answer your question and why?

3. What is your argument?

Your report should answer these questions in two pages or less. You should write your report using a word processing application (such as Word, Pages, or Google Docs), if possible, to make it easier for you to edit and revise it later. You should embed any diagrams, figures, or tables into the document. Be sure to write in a persuasive style; you are trying to convince others that your claim is acceptable or valid.

References

National Weather Service. n.d. Moore, Oklahoma tornadoes (1890–present). *www.weather.gov/oun/tornadodata-city-ok-moore.*

National Weather Service. 2017. Tornadoes in the Oklahoma City, Oklahoma area since 1890. *www.weather.gov/oun/tornadodata-okc.*

Thompson, A. 2013. New satellite image shows Moore tornado scar. *www.livescience.com/37176-moore-tornado-damage-satellite-image.html.*

Checkout Questions

Lab 21. Forecasting Extreme Weather: When and Under What Atmospheric Conditions Are Tornadoes Likely to Develop in the Oklahoma City Area?

1. What atmospheric conditions are typically present before tornado formation?

2. The table below shows data for four moments in Oklahoma City with different atmospheric conditions.

Moment	Date	Time	Atmospheric conditions
A	January 15	5:00–7:00 p.m.	Approaching warm front
B	May 3	5:00–7:00 p.m.	Approaching cold front
C	January 15	7:00–9:00 a.m.	Approaching cold front
D	May 3	7:00–9:00 a.m.	Stationary front

a. Rank the moments in the order of likelihood that a severe tornado will occur during this time, with 1 being most likely that a severe tornado will occur, and 4 being least likely that a severe tornado will occur.

b. How do you know?

LAB 21

3. A list of every date a tornado has struck the Oklahoma City area is an example of evidence.

    a. I agree with this statement.

    b. I disagree with this statement.

Explain your answer, using an example from your investigation about forecasting extreme weather.

4. Scientists do not always use lab experiments to further a scientific understanding of a natural phenomenon.

    a. I agree with this statement.

    b. I disagree with this statement.

Explain your answer, using an example from your investigation about forecasting extreme weather.

5. Scientists often need to look for patterns that occur in the data they collect and analyze. Explain why identifying patterns is important, using an example from your investigation about forecasting extreme weather.

6. Natural phenomena have causes, and uncovering causal relationships is a major activity of science. Explain why it is important to uncover causal relationships, using an example from your investigation about forecasting extreme weather.

Application Labs

Lab Handout

Lab 22. Minimizing Carbon Emissions: What Type of Greenhouse Gas Emission Reduction Policy Will Different Regions of the World Need to Adopt to Prevent the Average Global Surface Temperature on Earth From Increasing by 2°C Between Now and the Year 2100?

Introduction

When sunlight first reaches Earth, some of it is either reflected back out into space or absorbed by the atmosphere. The rest of the sunlight travels to Earth's surface. The sunlight that travels through the atmosphere hits Earth's surface, and the energy from the sunlight increases the temperature of Earth's surface.

Earth's surface emits infrared radiation, and the amount of infrared radiation emitted by the surface increases at higher temperatures. The atmosphere traps some of this infrared radiation before it can escape into space. The trapped infrared radiation in the atmosphere helps keep the temperature of Earth warmer than it would be without the atmosphere. Scientists call the warming of the atmosphere that is caused by trapped infrared radiation the greenhouse effect. Many gases that are found naturally in Earth's atmosphere, including water vapor, carbon dioxide (CO_2), methane, nitrous oxide, and ozone, are called greenhouse gases because these gases trap infrared energy in the atmosphere (see Figure L22.1).

The amount of energy that enters and leaves the Earth system is directly related to the average global temperature of the Earth. The Earth system, which includes the surface and the atmosphere, currently absorbs an average of about 240 watts of solar power per square meter over the course of the year. The Earth system also emits about the same amount of infrared energy into space. The average global surface temperature, as a result, tends to be stable over time. However, if something was to change the amount of energy that enters or leaves this system, then the flow of energy would be unbalanced and the average global temperature would change in response. Therefore, any change to the Earth system that affects how much energy enters or leaves the system can cause a significant change in Earth's average global temperature.

Climate scientists agree that human activities have led to an increase in the concentration of greenhouse gases found in the atmosphere (IPCC 2013). The greenhouse gas that has increased the most is carbon dioxide. Plants and animals release CO_2 into the atmosphere as a waste product of respiration. Plants also absorb CO_2 from the atmosphere for

Minimizing Carbon Emissions

What Type of Greenhouse Gas Emission Reduction Policy Will Different Regions of the World Need to Adopt to Prevent the Average Global Surface Temperature on Earth From Increasing by 2°C Between Now and the Year 2100?

FIGURE L22.1

The greenhouse effect

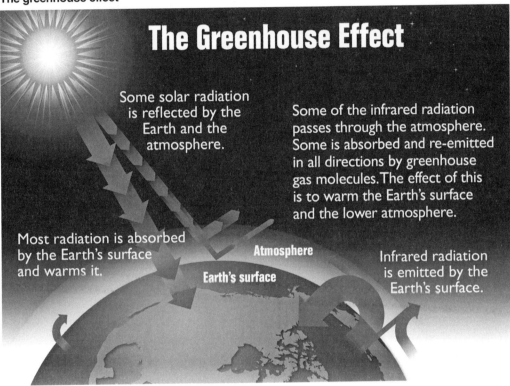

photosynthesis. People, however, also add CO_2 to the atmosphere when they burn fossil fuels for energy. This energy is used for electricity; as fuel for cars, planes, or boats; for heating and cooling homes; and for industrial manufacturing. The concentration of CO_2 in the atmosphere increased from about 320 parts per million (ppm) to almost 390 ppm between 1960 and 2010 (see Figure L22.2). The increase in greenhouse gas emissions such as CO_2 has led to an increase in the greenhouse effect of the atmosphere. The increased greenhouse effect has caused the average global surface temperature to increase approximately 0.8°C (1.4°F) since 1880.

FIGURE L22.2

Concentration of carbon dioxide in the atmosphere between 1960 and 2010; measurements recorded at Mauna Loa, Hawaii (the levels at this site are considered representative of global carbon dioxide levels)

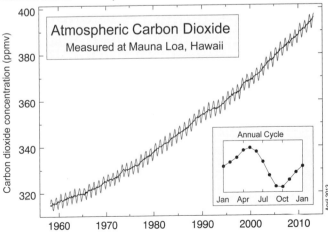

Greenhouse gas concentrations in the atmosphere will continue to increase unless the billions of tons of greenhouse gas that people add to the atmosphere each year decreases substantially. Even if greenhouse emissions were to stop completely, atmospheric greenhouse gas concentrations will continue to remain elevated for hundreds of years because many greenhouse gases stay in the atmosphere for long periods of time. Moreover, if the concentration of greenhouse gases found in the current atmosphere remained steady (which would require a dramatic reduction in current greenhouse gas emissions), surface air temperatures will continue to increase. This is because the oceans, which store heat, take many decades to fully respond to changes in greenhouse gas concentrations. The oceans will therefore continue to have an impact on global climate over the next several decades to hundreds of years.

The Intergovernmental Panel on Climate Change (IPCC), which is the leading international body for the assessment of change in global climate, expects the Earth's average temperature to increase by at least 4.5°C (8.1°F) by the year 2100 (IPCC 2013). This substantial increase in average global temperature will reduce ice and snow cover, change precipitation patterns across the globe, raise sea level, increase the acidity of the oceans, and shift the characteristics of many different natural habitats. These changes will have an impact on our food supply, water resources, infrastructure, and health. Many people are therefore examining different ways to reduce the amount of CO_2 that is released into the atmosphere or to find ways to remove more CO_2 from the atmosphere to help prevent this substantial increase in average global temperature. The current goal is for various regions of the world to identify and enact CO_2 emission reduction policies that will prevent the average global surface temperature from increasing by more than 2°C (3.6°F) before the year 2100.

Your Task

Use what you know about the greenhouse effect, systems and system models, and stability and change to develop a plan to minimize human impact on climate. Your goal is to identify a greenhouse gas emission reduction plan for six different regions of the world that will, when enacted together, prevent the average global surface temperature of Earth from increasing by more than 2°C until the year 2100. The plan you will develop for each region must include recommendations for when the greenhouse gas emissions in the region will need to stop increasing, when the greenhouse gas emissions will need to start decreasing in that region, and the annual greenhouse gas emission reduction rate for that the region. Each plan will also need to include recommendations for a policy that will dictate how much deforestation will need to be reduced in that region by 2050 and how much effort should go into creating new forests in that region starting in 2016.

The guiding question of this investigation is, *What type of greenhouse gas emission reduction policy will different regions of the world need to adopt to prevent the average global surface temperature on Earth from increasing by 2°C between now and the year 2100?*

Minimizing Carbon Emissions

What Type of Greenhouse Gas Emission Reduction Policy Will Different Regions of the World Need to Adopt to Prevent the Average Global Surface Temperature on Earth From Increasing by 2°C Between Now and the Year 2100?

Materials

You will use an online simulation called C-ROADS (Climate Rapid Overview and Decision Support) that helps people understand the long-term climate impacts of actions that reduce greenhouse gas emissions; the simulation is available at *www.climateinteractive.org/tools/c-roads*.

Safety Precautions

Follow all normal lab safety rules.

Investigation Proposal Required?   ☐ Yes   ☐ No

Getting Started

You can use C-ROADS (see Figure L22.3) to test strategies for tackling climate change. The online simulation allows users to visualize the impact of adopting different emission reduction policies in different regions of the world. This model is useful because it allows users to see and understand the gap between a proposed policy (such as reducing deforestation in a region by 10% a year) and what actually needs to happen to stabilize the concentration of greenhouse gases in the atmosphere and prevent an increase in average global surface temperature. C-ROADS therefore offers you a way to test different policies.

FIGURE L22.3 _____

Screenshot of C-ROADS simulation

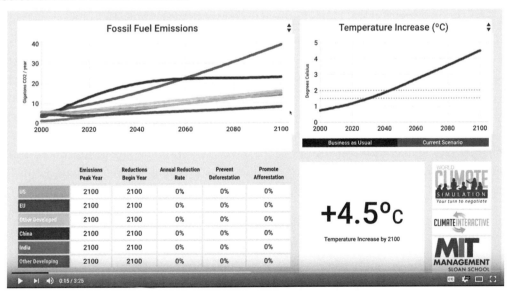

The first step in this investigation is to develop a greenhouse gas emission reduction policy for different regions of the world and then see how these policies, when enacted together, will affect the average global surface temperature over time. *C-ROADS* allows you to create greenhouse gas emission reduction policies for (1) the United States, (2) the European Union, (3) Other Developed Countries (e.g., Japan, Russia, Australia, Korea), (4) China, (5) India, and (6) Other Developing Countries (e.g., nations in Africa, South America, Central America, and the Middle East). The policy you create for each region will need to include

- the year that greenhouse gas emissions will *stop increasing* in the region (emission peak year),
- the year that greenhouse gas emission will *start decreasing* in the region (reduction begin year),
- the rate at which each region *decreases* emission each year (annual reduction rate),
- the rate at which deforestation will *decrease* in the region each year (reduce deforestation), and
- the rate at which new forests will *added* in the region each year (promote afforestation).

You can set values for these five aspects of a greenhouse gas emission reduction policy for the six different regions in *C-ROADS* using the table in the lower-left corner of the simulation. You can then examine how your choices will affect the fossil fuel emissions in each region over time (graph in the upper-left corner) and the average global temperature over time (graph in the upper-right corner). The box in the lower-right corner also tells you how much the average global temperature will change between now and 2100.

One you have developed an initial policy for each region and tested how these policies, when enacted together, will affect fossil fuel emissions and average global temperature over time, you can then begin to modify each of your proposed policies to make your overall plan more effective. Your goal for this step of the investigation is to determine how your proposed policies for each region will need to be changed to close the gap between what you are proposing and what actually needs to happen to prevent a 2°C increase in average global surface temperature between now and 2100. As you modify your greenhouse gas emission policies, think about the following questions to help guide the changes you will make:

- What are the components of the system and how do they interact with each other?
- When is this system stable and under which conditions does it change?
- Which factor(s) might control the rate of change in this system?
- What scale or scales should you use to when you take measurements?

The last step in your investigation will be to identify the potential challenges and consequences that are associated with putting a proposed greenhouse gas emission reduction policy into action. This is important because policies affect the lives of people. For example, what are the consequences of a policy that requires an immediate 20% reduction in fossil fuel emissions for the people in a region that relies on coal or oil as an energy source? Will the people in that region be able to get to work? Heat their homes? Will they lose their jobs? You must also consider what is ethical and what is fair in terms of any proposed policy. For example, is it ethical or fair for people who live in a country in South America to have to cut greenhouse gas emissions at the same rate as people who live in the United States? You will therefore need to take into account these issues to ensure that your proposed greenhouse gas emission reduction policy is valid and acceptable. You can include information about any political, economic, or social issues you considered as you developed and evaluated your policy as part of your justification of your evidence.

Connections to the Nature of Scientific Knowledge and Scientific Inquiry

As you work through your investigation, be sure to think about

- how models are used as tools for reasoning about natural phenomena, and
- the types of questions that scientists can investigate.

Initial Argument

Once your group has finished collecting and analyzing your data, your group will need to develop an initial argument. Your initial argument needs to include a claim, evidence to support your claim, and a justification of the evidence. The *claim* is your group's answer to the guiding question. The *evidence* is an analysis and interpretation of your data. Finally, the *justification* of the evidence is why your group thinks the evidence matters. The justification of the evidence is important because scientists can use different kinds of evidence to support their claims. Your group will create your initial argument on a whiteboard. Your whiteboard should include all the information shown in Figure L22.4.

FIGURE L22.4
Argument presentation on a whiteboard

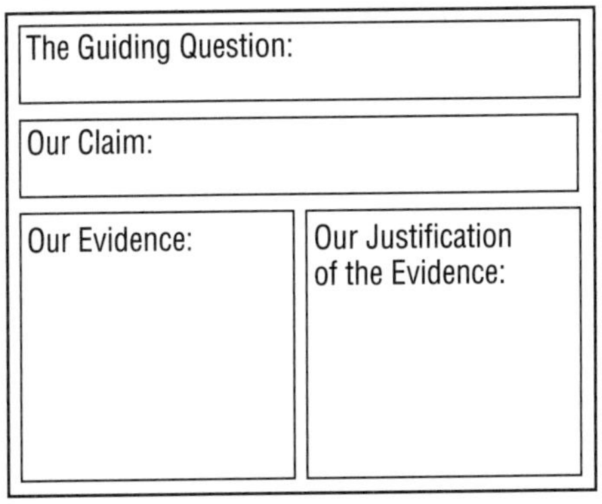

The Guiding Question:

Our Claim:

Our Evidence: | Our Justification of the Evidence:

Argumentation Session

The argumentation session allows all of the groups to share their arguments. One or two members of each group will stay at the lab station to share that group's argument, while the other members of the group go to the other lab stations to listen to and critique the other arguments. This is similar to what scientists do when they propose, support, evaluate, and refine new ideas during a poster session at a conference. If you are presenting your group's argument, your goal is to share your ideas and answer questions. You should also keep a record of the critiques and suggestions made by your classmates so you can use this feedback to make your initial argument stronger. You can keep track of specific critiques and suggestions for improvement that your classmates mention in the space below.

Critiques of our initial argument and suggestions for improvement:

If you are critiquing your classmates' arguments, your goal is to look for mistakes in their arguments and offer suggestions for improvement so these mistakes can be fixed. You should look for ways to make your initial argument stronger by looking for things that the other groups did well. You can keep track of interesting ideas that you see and hear during the argumentation in the space below. You can also use this space to keep track of any questions that you will need to discuss with your team.

Interesting ideas from other groups or questions to take back to my group:

Minimizing Carbon Emissions

What Type of Greenhouse Gas Emission Reduction Policy Will Different Regions of the World Need to Adopt to Prevent the Average Global Surface Temperature on Earth From Increasing by 2°C Between Now and the Year 2100?

Once the argumentation session is complete, you will have a chance to meet with your group and revise your initial argument. Your group might need to gather more data or design a way to test one or more alternative claims as part of this process. Remember, your goal at this stage of the investigation is to develop the best argument possible.

Report

Once you have completed your research, you will need to prepare an *investigation report* that consists of three sections. Each section should provide an answer for the following questions:

1. What question were you trying to answer and why?

2. What did you do to answer your question and why?

3. What is your argument?

Your report should answer these questions in two pages or less. You should write your report using a word processing application (such as Word, Pages, or Google Docs), if possible, to make it easier for you to edit and revise it later. You should embed any diagrams, figures, or tables into the document. Be sure to write in a persuasive style; you are trying to convince others that your claim is acceptable or valid.

Reference

Intergovernmental Panel on Climate Change (IPCC). 2013. *Climate change 2013: The physical science basis. Working Group I contribution to the Fifth Assessment Report of the Intergovernmental Panel on Climate Change.* [Stocker, T.F., D. Qin, G.-K. Plattner, M. Tignor, S.K. Allen, J. Boschung, A. Nauels, Y. Xia, V. Bex and P.M. Midgley (eds.)]. Cambridge, England: Cambridge University Press.

Checkout Questions

Lab 22. Minimizing Carbon Emissions: What Type of Greenhouse Gas Emission Reduction Policy Will Different Regions of the World Need to Adopt to Prevent the Average Global Surface Temperature on Earth From Increasing by 2°C Between Now and the Year 2100?

1. In the past century, how have humans increased the amount of carbon dioxide in the atmosphere?

2. What methods can produce usable energy without carbon dioxide emissions?

Minimizing Carbon Emissions

What Type of Greenhouse Gas Emission Reduction Policy Will Different Regions of the World Need to Adopt to Prevent the Average Global Surface Temperature on Earth From Increasing by 2°C Between Now and the Year 2100?

3. A student has submitted the following plan to reduce carbon dioxide emissions to 20% of the levels in 1990.

Action area	Policy	Result
Biofuel production	Government subsidies on land used for biofuel	Use 8,000 square miles of land to grow biofuels. Additionally, import the same amount of biofuels.
Oil, gas, and coal power	Some oil refineries and coal plants shut down to reduce availability of fossil fuels	Reduce the amount of fossil fuels used by 50%
Nuclear power	Tax reductions on nuclear power plants	Build 13 large nuclear power plants around the country.
Wind turbines	Tax credits for businesses that install wind turbines; state funding allocated for building wind turbines	Build 13,000 wind turbines on land and 17,000 wind turbines offshore.
Manufacturing growth	Strict regulations on new and existing manufacturing companies	Manufacturing declines, becoming roughly one-third smaller than existing manufacturing.
Home efficiency	Tax breaks and refunds for adding insulation to new and existing homes	Additional installation is installed in 75% of homes.
Home temperature	Radio, television, and internet advertisements that encourage thermostat changes	The average home temperature in the winter decreases from 17.5°C to 17°C.
Heating fuel	Increased taxes on coal- or gas-burning furnaces	About 20% of domestic heat is powered by electricity.
How we travel	Decreased prices on public transportation, as well as an increase in number of stops	People use public transportation for about one-quarter of their journeys.
Transport fuel	Government regulation that three out of five of the cars sold must be powered by electricity	Three out of five of cars driven are powered by electricity.

a. Which actions seem feasible for a country to carry out by 2050?

b. Why?

c. Which actions do not seem feasible for a country to carry out by 2050?

d. Why not?

4. Models are pictures created by scientists to teach something to others.

   a. I agree with this statement.
   b. I disagree with this statement.

   Explain your answer, using an example from your investigation about minimizing carbon emissions.

Minimizing Carbon Emissions

What Type of Greenhouse Gas Emission Reduction Policy Will Different Regions of the World Need to Adopt to Prevent the Average Global Surface Temperature on Earth From Increasing by 2°C Between Now and the Year 2100?

5. Science can answer any question.

   a. I agree with this statement.

   b. I disagree with this statement.

   Explain your answer, using an example from your investigation about minimizing carbon emissions.

6. Defining a system under study and making a model of it are tools for developing a better understanding of natural phenomena in science. Explain why it is important to make models of natural phenomena, using an example from your investigation about minimizing carbon emissions.

7. When studying a system, one of the main objectives is to determine how the system is changing over time and which factors are causing the system to become unstable. Explain why determining how a system changes over time is important, using an example from your investigation about minimizing carbon emissions.

Lab Handout

Lab 23. Human Use of Natural Resources: Which Combination of Water Use Policies Will Ensure That the Phoenix Metropolitan Area Water Supply Is Sustainable?

Introduction

Water is an essential resource for us. We must drink water to survive. People also use water to cook, to clean, and for recreational purposes. We also need water to grow the food that we need and to produce many of the products that we use on a daily basis. Many people think water is an unlimited resource because oceans cover about 70% of Earth's surface. The water found in oceans, however, is high in salt and not fit for human consumption. In fact, 97% of the water found on Earth is classified as salt water (see Figure L23.1), so only 3% of all the water found on Earth is fresh. Of all this freshwater, about 69% of it is frozen in glaciers and the ice caps, 30% is groundwater, and the remaining 1% is located on the surface in lakes, rivers, marshes,

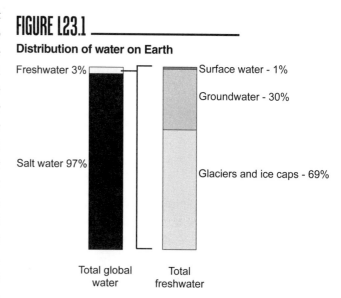

FIGURE L23.1

Distribution of water on Earth

Freshwater 3%

Salt water 97%

Surface water - 1%

Groundwater - 30%

Glaciers and ice caps - 69%

Total global water

Total freshwater

and swamps. All our drinking water and the water we use for agriculture, manufacturing, and sanitation comes from this relatively small amount of surface water and groundwater.

Water reservoirs include natural or human-made lakes (see Figure L23.2) and the groundwater that is found in large aquifers, which are underground stores of freshwater (see Figure L23.3). Scientists use the term *recharge* to describe how a reservoir fills with water over time. The rate of recharge depends on the amount of precipitation that happens in an area. Many people depend on an aboveground and/or belowground water reservoir to supply all the water they use on a daily basis. Unfortunately, a water reservoir can be depleted of water faster than it can recharge when people consume too much water and there is not much precipitation in an area for an extended amount of time. When a water reservoir runs dry, people who live in that area are forced to do without the water they need until it fills again. High consumption of water can also have a negative impact on the local environment. Typically, as human population and per capita (per person) consumption of water in a region increases, so does the likelihood that the people and the environment in that region will experience a negative impact. It is therefore important for people in a region to find a way to use water in a sustainable manner.

FIGURE L23.2

The Lake Mead Reservoir in Nevada and Arizona

FIGURE L23.3

Illustration of an aquifer

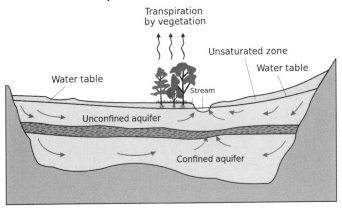

High hydraulic-conductivity aquifer

Low hydraulic-conductivity confining unit

Very low hydraulic-conductivity bedrock

Direction of ground-water flow

One way that a group of people can ensure that their use of water over time is sustainable is for them to establish and then follow policies that will always keep the amount or rate of water use (*demand*) at a level that is equal to or below the amount of water that is or will be available in the local water reservoir (*supply*). The state of Arizona is a good example of how people can work together to ensure that their consumption of water is sustainable over time despite having a limited supply

Arizona is one of the driest states in the United States. In fact, it only receives a statewide average of 12.5 inches of rain per year. Arizona is also one of the fastest-growing states in terms of population. Arizona's population in 2010 was 6.4 million and is projected to increase to over 9.5 million people by 2025 (U.S. Census Bureau 2012). Most of the people who live in Arizona reside within the Phoenix metropolitan area (PMA). The population in the PMA in 2010 was about 4.2 million (U.S. Census Bureau 2012). This many people living in the same area can use a lot of water over the course of a year. In fact, people living in the PMA used 3,667 acre-feet (1 acre-foot = 325,851 gallons) of water in 2008 (Arizona

Department of Water Resources [2009]). The dry climate presents numerous challenges for the people living in Arizona because they can quickly deplete their limited water supply. Therefore, the people of Arizona must find sustainable ways to use their limited supply of water to ensure that they can maintain their quality of life and grow their economy without damaging the local environment.

The people of Arizona get the water they need for agriculture, industry, and municipal use from three major sources. *The first major source is surface water,* and the largest portion of surface water comes from two main reservoirs on the Colorado River (Lake Mead and Lake Powell). The Colorado River starts in the central Rocky Mountains and drains into the Gulf of California (see Figure L23.4). The Colorado River supplies water to people in Arizona, California, Nevada, New Mexico, Utah, Colorado, Wyoming, and Mexico.

FIGURE L23.4

A map of the Colorado River watershed

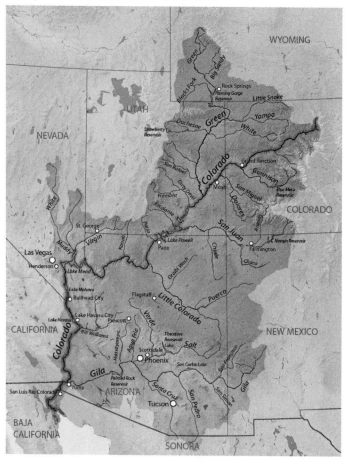

The people of Arizona are only allowed to take 2.8 million acre-feet of water from the Colorado River reservoirs annually (Arizona Department of Water Resources n.d.).

Smaller rivers and lakes in Arizona also provide some water for the people of Arizona. However, the amount of water available from these sites varies from year to year, season to season, and place to place because of the desert climate. The Arizona government has therefore built reservoir storage systems in most of the major rivers within the state, such as the Salt, Verde, Gila, and Agua Fria.

The second major source of water is groundwater. Aquifers supply about 43% of the state's water (Arizona Department of Water Resources n.d.). Throughout the 20th and 21st centuries, however, groundwater has been pumped out of the aquifers faster than it could recharge, which has left them depleted. Though a large amount of water remains stored underground, its availability is limited.

The third major source of water in Arizona is effluent. Effluent, or reclaimed water, is wastewater that has been collected and treated. The people of Arizona use reclaimed water for agriculture, golf courses, industrial cooling, and to maintain parks and other wildlife areas.

To help the people of the PMA identify and enact policies that will ensure that the available water supply is used in a sustainable manner, scientists at Arizona State University created a visualization tool called *WaterSim. WaterSim* uses a mathematical model to estimate water supply and demand for the PMA. People can use the *WaterSim* visualization tool to explore how various regional population growth, drought, and climate change scenarios and water management policies affect water sustainability. This visualization tool is valuable because it takes a lot of data that are usually collected separately (including water supply, water demand, climate, population, and policy data) and puts them together to give the user a way to see how all these different variables interact with each other. It also allows a user to change one variable in the system at a time and see how that change affects the other components of the system. In this investigation, you will have an opportunity to use the *WaterSim* visualization tool to explore different water use policies to determine how these different policies, if enacted, will affect the sustainability of the PMA water supply.

Your Task

Use what you know about natural resources, human impacts on Earth systems, stability and change, and cause-and-effect relationships to determine how different water use policies will affect the current and near-future needs of the people living in the PMA. To accomplish this goal, you will need to use the *WaterSim* visualization tool to examine how different water use choices will affect the water supply between now and the year 2050, assuming best- and worst-case scenarios for population growth, drought, and climate change. Your analysis, at a minimum, will need to examine per capita water use, the percentage of wastewater to be reclaimed, and the percentage of farm water to be used by cities in the PMA. It should also include an analysis of how much of the water in the Colorado River can be used to help restore the Colorado River delta of northern Mexico to prevent future habitat and biodiversity loss.

The guiding question of this investigation is, **Which combination of water use policies will ensure that the Phoenix Metropolitan Area water supply is sustainable?**

Materials

You will use a visualization tool called WaterSim to explore how water policy decisions influence water supply and sustainability; the tool is available at *https://sustainability.asu.edu/dcdc/watersim.*

Safety Precautions

Follow all normal lab safety rules.

LAB 23

Investigation Proposal Required? ☐ Yes ☐ No

Getting Started

You can use the *WaterSim* visualization tool (see Figure L23.5) to examine how different water use choices affect the sustainability of water in the PMA over time, assuming different conditions in Arizona. This tool is useful because it allows users to see and understand the gap between a proposed policy choice (such as the percentage of wastewater to be reclaimed or how much farm water should be diverted to cities) and what actually needs to happen to ensure the sustainability of a water supply. *WaterSim* therefore offers you a way to test different policies.

The first step in this investigation is to develop an overall water use plan for the PMA and then see how this plan will affect the sustainability of water over time. *WaterSim* allows you to create a water use plan for the PMA that includes the following policy choices:

- the amount of total wastewater from residential, commercial, and industrial water users that is diverted to the reclaimed wastewater treatment plant ("% of Wastewater Reclaimed");
- the agriculture (farming) water made available for urban water use ("Farm Water Used by Cities");
- the portion of the Colorado River water flow intentionally left in the river to be diverted for use by the Colorado River delta ("Environmental Flows"); and
- the amount of water that people use at home, which is measured in units of *gallons per capita per day* (GPCD) per person ("Per Capita Water Use")

You can set values for these four water use policy choices using the sliders in the lower-left corner of the simulation.

Next, you can examine how these choices will affect the sustainability of the PMA water system using five different indicators at the top of the simulation. Each indicator represents a different aspect of water sustainability:

- *Groundwater.* How much groundwater needs to be extracted from the aquifers to meet the demands of the community as a percentage of the total amount of water use.
- *Environment.* The percentage contribution of Arizona's commitment to leave water in the Colorado River to restore the Colorado River delta
- *Ag to Urban.* The amount of agriculture water credits being diverted from farming and agriculture for municipal water use. Agriculture is a major part of the economy in Arizona, so less water available for agriculture means less economic growth.
- *Personal.* The amount of water that can be used for people to cook, clean, bathe, dispose of waste, and irrigate yards. Water is needed for health, comfort,

and overall well-being, so less water available for personal use means more inconveniences for people.

- *Population.* How many years the water supply can support the current population before the population would experience water deficits or, alternatively, have to find more water at a higher cost.

The value presented inside the box for each indicator represents the final value (end of 2050) over the simulation period. The values for each indicator from the previous simulation are retained at the bottom of each indicator (in parentheses).

FIGURE L23.5

Screenshot of the *WaterSim* visualization tool

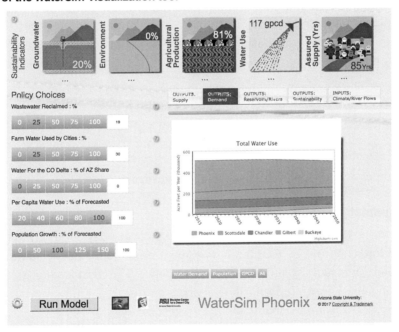

Once you have identified an overall water use plan for the PMA and determined how it will affect water sustainability over time, you can then begin to modify each water use policy choice to make the overall plan more effective. You can also change how fast the population will grow over time, how often and how long droughts will occur, and how the climate may change over time in the *WaterSim* visualization tool. Your goal for this step of the investigation is to determine how different policies will need to be modified to close the gap between what is proposed and what actually needs to happen to ensure the sustainability of the PMA water supply. As you modify the different water use policies and your assumptions about population growth, droughts, and climate change in the *WaterSim* visualization tool, think about the following questions to help guide the changes you will make:

- What are the components of the system and how do they interact with each other?
- When is this system stable and under which conditions does it change?
- Which factor(s) might control the rate of change in this system?
- What scale or scales should you use when you take measurements?

The last step in your investigation will be to identify the potential challenges and consequences that are associated with putting a proposed water use plan into action. This is important because policies affect the lives of people. For example, what are the consequences to farmers of a policy that requires more water to be diverted from farms to cities? Will farmers lose their jobs or their land? What will this do to the overall economy? You must also consider what is ethical and what is fair in terms of any proposed policy. For example, does a policy impact the people who live in cities and in rural areas the same way? You will need to take these issues into account to ensure that your proposed water use plan is valid and acceptable. You can include information about any political, economic, or social issues you considered as you developed and evaluated your policy as part of your justification of your evidence.

Connections to the Nature of Scientific Knowledge and Scientific Inquiry

As you work through your investigation, be sure to think about

- how scientific knowledge can change over time, and
- the types of questions that scientists can investigate.

Initial Argument

Once your group has finished collecting and analyzing your data, your group will need to develop an initial argument. Your initial argument needs to include a claim, evidence to support your claim, and a justification of the evidence. The *claim* is your group's answer to the guiding question. The *evidence* is an analysis and interpretation of your data. Finally, the *justification* of the evidence is why your group thinks the evidence matters. The justification of the evidence is important because scientists can use different kinds of evidence to support their claims. Your group will create your initial argument on a whiteboard. Your whiteboard should include all the information shown in Figure L23.6.

FIGURE L23.6

Argument presentation on a whiteboard

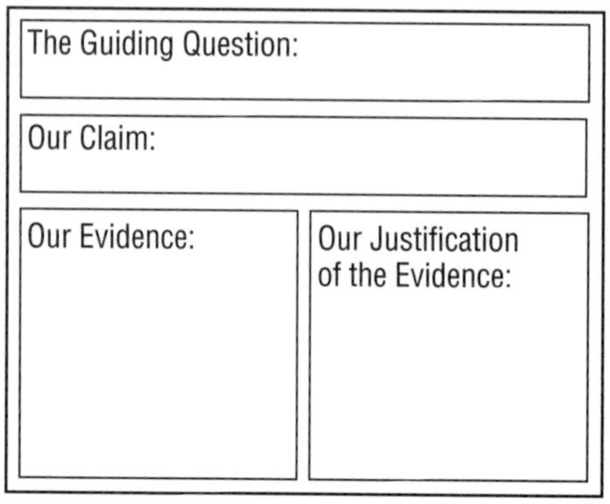

The Guiding Question:	
Our Claim:	
Our Evidence:	Our Justification of the Evidence:

Argumentation Session

The argumentation session allows all of the groups to share their arguments. One or two members of each group will stay at the lab station to share that group's argument, while the other members of the group go to the other lab stations to listen to and critique the other arguments. This is similar to what scientists do when they propose, support, evaluate, and refine new ideas during a poster session at a conference. If you are presenting your group's argument, your goal is to share your ideas and answer questions. You should also keep a record of the critiques and suggestions made by your classmates so you can use this feedback to make your initial argument stronger. You can keep track of specific critiques and suggestions for improvement that your classmates mention in the space below.

Critiques of our initial argument and suggestions for improvement:

If you are critiquing your classmates' arguments, your goal is to look for mistakes in their arguments and offer suggestions for improvement so these mistakes can be fixed. You should look for ways to make your initial argument stronger by looking for things that the other groups did well. You can keep track of interesting ideas that you see and hear during the argumentation in the space below. You can also use this space to keep track of any questions that you will need to discuss with your team.

Interesting ideas from other groups or questions to take back to my group:

Once the argumentation session is complete, you will have a chance to meet with your group and revise your initial argument. Your group might need to gather more data or design a way to test one or more alternative claims as part of this process. Remember, your goal at this stage of the investigation is to develop the best argument possible.

Report

Once you have completed your research, you will need to prepare an *investigation report* that consists of three sections. Each section should provide an answer for the following questions:

1. What question were you trying to answer and why?

2. What did you do to answer your question and why?

3. What is your argument?

Your report should answer these questions in two pages or less. You should write your report using a word processing application (such as Word, Pages, or Google Docs), if possible, to make it easier for you to edit and revise it later. You should embed any diagrams, figures, or tables into the document. Be sure to write in a persuasive style; you are trying to convince others that your claim is acceptable or valid.

References

U.S. Census Bureau. 2012. Arizona: 2010. Population and housing unit counts. 2010 Census of population and housing. CPH-2-4. Washington, DC: U.S. Government Printing Office. Also available online at *www.census.gov/prod/cen2010/cph-2-4.pdf.*

Arizona Department of Water Resources. n.d. Securing Arizona's water future. *www.azwater.gov/AzDWR/PublicInformationOfficer/documents/supplydemand.pdf.*

Arizona Department of Water Resources Drought Program. [2009]. Community water systems 2008 annual water use reporting summary. *www.azwater.gov/AzDWR/StatewidePlanning/drought/2008AnnualWaterUse.htm.*

Checkout Questions

Lab 23. Human Use of Natural Resources: Which Combination of Water Use Policies Will Ensure That the Phoenix Metropolitan Area Water Supply Is Sustainable?

1. List two ways increasing human population affects the supply and sustainability of groundwater.

2. Three city plans for water use are listed in the table below. Assume that the population size and per capita use are 100% as forecasted.

City	Percent wastewater reclaimed	Percent farm water used by cities	Percent environmental flows (water that remains unused in river)
A	82	58	46
B	47	22	62
C	84	18	41

a. Which city plan will result in the most sustainable groundwater stores?

b. How do you know?

c. Which city plan will result in the least sustainable groundwater stores?

d. How do you know?

3. Scientific explanations are subject to revision and improvement in light of new evidence.

   a. I agree with this statement.
   b. I disagree with this statement.

   Explain your answer, using an example from your investigation about human use of natural resources.

4. Science can answer all questions.

   a. I agree with this statement.
   b. I disagree with this statement.

   Explain your answer, using an example from your investigation about human use of natural resources.

5. One of the main objectives of science is to identify and establish relationships between a cause and an effect. Explain why identifying cause-and-effect relationships is important, using an example from your investigation about human use of natural resources.

6. Explain why it is important to determine how a system is changing over time and which factors are causing the system to become unstable, using an example from your investigation about human use of natural resources.

IMAGE CREDITS

All images in this book are stock photographs or courtesy of the authors unless otherwise noted below.

Lab 1

Figure L1.1: (Waxing Crescent) Jay Tanner, Wikimedia Commons, CC BY-SA 3.0, *https:// upload.wikimedia.org/wikipedia/commons/7/7e/ Phase-048.jpg*; (First Quarter) Jay Tanner, Wikimedia Commons, CC BY-SA 3.0, *https://upload. wikimedia.org/wikipedia/commons/c/c0/Phase-098. jpg*; (Waxing Gibbous) Jay Tanner, Wikimedia Commons, CC BY-SA 3.0, *https://upload. wikimedia.org/wikipedia/commons/6/64/Phase-133. jpg;* (Full) Jay Tanner, Wikimedia Commons, CC BY-SA 3.0, *https://upload.wikimedia.org/wikipedia/ commons/0/01/Phase-191.jpg*; (Waning Gibbous) Jay Tanner, Wikimedia Commons, CC BY-SA 3.0, *https://upload.wikimedia.org/wikipedia/ commons/3/34/Phase-231.jpg*; (Third Quarter) Jay Tanner, Wikimedia Commons, CC BY-SA 3.0, *https:// upload.wikimedia.org/wikipedia/commons/6/6e/ Phase-270.jpg*; (Waning Crescent) Jay Tanner, Wikimedia Commons, CC BY-SA 3.0, *https://upload. wikimedia.org/wikipedia/commons/e/ef/Phase-312. jpg*; (New) Jay Tanner, Wikimedia Commons, CC BY-SA 3.0, *https://upload.wikimedia.org/wikipedia/ commons/3/36/Phase-358.jpg*

Figure L1.2: SBS, "Solar Eclipse: Myths and Facts," July 23, 2009, *www.sbs.com.au/news/ article/2009/07/22/solar-eclipse-myths-and-facts*

Figure L1.3: Tom Ruen, Wikimedia Commons, Public domain. *http://commons.wikimedia.org/wiki/ File:Partial_lunar_eclipse_december_10_2011_ Minneapolis_TLR.png*

Lab 2

Figure L2.1: Jet Propulsion Laboratory, NASA, "Season's Greetings: NASA Views the Change of Seasons," December 21, 2010, *www.jpl.nasa.gov/ news/news.php?feature=2857*

Figure L2.2: Dennis Nilsson, Wikimedia Commons, CC 3.0, *https://en.wikipedia.org/wiki/Axial_tilt#/ media/File:AxialTiltObliquity.png*

Figure L2.3: University of Nebraska–Lincoln, NAAP Astronomy Labs—Basic Coordinates and Seasons—Seasons and Ecliptic Simulator *http:// astro.unl.edu/naap/motion1/animations/seasons_ ecliptic.html*

Figure in checkout question 2: Modified from Rhcastilhos, Wikimedia Commons, Public domain. *https://upload.wikimedia.org/wikipedia/commons/ thumb/1/12/Seasons.svg/2000px-Seasons.svg.png*

Lab 3

Figure L3.4: NASA, Public domain. *http:// solarsystem.nasa.gov/galleries/earths-orbit*

Figure L3.5: PhET Interactive Simulations, Univeristy of Colorado—Bolder, *https://phet.colorado.edu/en/ simulation/legacy/my-solar-system*

Lab 5

Figure L5.1: (a) Luca Galuzzi, Wikimedia Commons, CC BY-SA 2.5, *https://commons.wikimedia.org/ wiki/File:USA_10052_Grand_Canyon_Luca_ Galuzzi_2007.jpg*; (b) Ryan Lackey, Wikimedia Commons, CC BY 2.0, *https://commons.wikimedia. org/wiki/File:Sedimentary_Rock_Layers_near_ Khasab_in_Musandam_Oman.jpg*

Figure L5.2: Peter Halasz, Wikimedia Commons, Public domain. *https://upload.wikimedia.org/ wikipedia/commons/thumb/a/a5/Biological_ classification_L_Pengo_vflip.svg/399px- Biological_classification_L_Pengo_vflip.svg.png*

Figure above checkout question 1: Kurt Rosenkrantz, Wikimedia Commons, CC BY-SA 3.0, *https://commons.wikimedia.org/wiki/File:Fossils.png*

Lab 6

Figure L6.1: U.S. Geological Survey, Wikimedia Commons, Public domain. *https://commons. wikimedia.org/wiki/File:FigS1-1.gif*

Figure L6.2: U.S. Geological Survey, Wikimedia Commons, Public domain. *https://en.wikipedia. org/wiki/Plate_tectonics#/media/File:Plates_tect2_ en.svg*

Figure above checkout question 1: U.S. Geological Survey, Wikimedia Commons, Public domain. *https:// en.wikipedia.org/wiki/Plate_tectonics#/media/ File:Plates_tect2_en.svg*

Image Credits

Figure in checkout question 3: Modified from Виктор, Wikimedia commons, CC BY-SA 2.0, *https://commons.wikimedia.org/wiki/File:Outline_map_of_Central_America.svg*

Lab 7

Figure L7.1: (a) NASA image courtesy Jeff Schmaltz, LANCE/EOSDIS MODIS Rapid Response Team at NASA GSFC, Wikimedia Commons, Public domain. *https://commons.wikimedia.org/wiki/File:Aleutian_Islands_amo_2014135_lrg*.jpg: (b) NASA, Wikimedia Commons, Public domain. *https://commons.wikimedia.org/wiki/File:Himalayas_landsat_7.png*

Figure L7.2: Surachit, Wikimedia Commons, GFDL 1.2, *https://en.wikipedia.org/wiki/Mantle_convection#/media/File:Oceanic_spreading.svg*

Figure in checkout question 2: Screenshot from Google Maps, *www.google.com/maps/place/Japan/@34.2521997,130.0725666,4z/data=!4m5!3m4!1s0x34674e0fd77f192f:0xf54275d47c665244!8m2!3d36.204824!4d138.252924*

Lab 8

Figure L8.1: El Guanche, Wikimedia Commons, CC BY 2.0, *https://en.wikipedia.org/wiki/Aeolian_processes#/media/File:Arbol_de_Piedra.jpg*

Figure L8.2: Roxy Lopez, Wikimedia Commons, CC BY-SA 3.0, *https://commons.wikimedia.org/wiki/File:Duststorm.jpg*

Figure L8.3: Natural Resources Conservation Service, U.S. Department of Agriculture, Public domain. *www.nrcs.usda.gov/wps/portal/nrcs/detail/soils/edu/?cid=nrcs142p2_054311*

Lab 9

Figure L9.1: National Park Service, Wikimedia Commons, Public domain. *https://commons.wikimedia.org/wiki/File:Mississippi_River_Watershed_Map.jpg*

Figure L9.2: NASA, Wikimedia Commons, Public domain. *https://en.wikipedia.org/wiki/Mississippi_River_Delta#/media/File:Mississippi_River_Delta_and_Sediment_Plume.jpg*

Figure above checkout question 1: Demis Map Server, Wikimedia Commons, CC BY-SA 3.0, *https://upload.wikimedia.org/wikipedia/commons/c/c2/Rogue_River_Watershed.png*

Lab 10

Figure L10.1: (Siltstone) Greg Willis, Wikimedia Commons, CC BY-SA 3.0, *https://commons.wikimedia.org/wiki/File:Balls_Bluff_Siltstone_(4802112634).jpg*; (Shale) Smurfage, Deviant Art, Public domain. *http://smurfage.deviantart.com/art/Shale-Wall-208582121*; (Sandstone) Óðinn, Wikimedia Commons, CC BY-SA 2.5, *https://commons.wikimedia.org/wiki/File:Navajo_Sandstone.JPG*

Lab 11

Figure L11.2: Natural Resources Conservation Service, U.S. Department of Agriculture, Public domain. *www.nrcs.usda.gov/wps/portal/nrcs/detail/soils/edu/?cid=nrcs142p2_054311*

Lab 12

Figure L12.1: (a) the_tahoe_guy, Wikimedia Commons, CC BY 2.0, *https://en.wikipedia.org/wiki/Lake_Tahoe#/media/File:Emerald_Bay.jpg*; (b) Luca Galuzzi, Wikimedia Commons, CC BY-SA 2.5, *https://en.wikipedia.org/wiki/Glacier#/media/File:Perito_Moreno_Glacier_Patagonia_Argentina_Luca_Galuzzi_2005.JPG*; (c) Diamonds [transliteration], Wikimedia Commons, CC BY_SA 3.0, *https://commons.wikimedia.org/wiki/File:Water_vapor_from_a_pond_in_Owakudani_Valley_2.JPG*

L12.2: Atmospheric Infrared Sounder, Flickr, CC BY 2.0, *www.flickr.com/photos/atmospheric-infrared-sounder/8265072146/sizes/l*

Lab 13

Figure L13.1: Jim Champion, Wikimedia Commons, GFDL 1.2, *https://commons.wikimedia.org/wiki/File:Logan_Rock_from_below.jpg*

Figure L13.2: Lucarelli, Wikimedia Commons, GFDL 1.2, *https://upload.wikimedia.org/wikipedia/commons/e/e3/Carrara_14.JPG*

Figure L13.3: Rob Lavinsky, Wikimedia Commons, CC BY-SA 3.0, *https://upload.wikimedia.org/wikipedia/commons/9/99/Dolomite-Calcite-201604.jpg*

L13.4: Didier Descouens, Wikimedia Commons, CC BY-SA 4.0, *https://upload.wikimedia.org/wikipedia/commons/c/ce/Quartz_Brésil.jpg*

Lab 14

Figure L14.1: Jonathan Zander, Wikimedia Commons, GFDL 1.2, *https://commons.wikimedia.org/wiki/File:Native_Copper_Macro_Digon3.jpg*

Figure L14.2: Matthew.kowal, Wikimedia Commons, CC BY-SA 4.0, *https://commons.wikimedia.org/wiki/File:The_Lavender_Open_Pit_Mine,_Bisbee,_Arizona.jpg*

Figure L14.3: Supercarwaar, Wikimedia Commons, CC BY-SA 4.0, *https://commons.wikimedia.org/wiki/File:Kennecott_Tailings_Pond_and_Smokestack.jpg*

Figure below checkout question 1: Kbh3rd, Wikimedia Commons, CC BY-SA 3.0, *https://commons.wikimedia.org/wiki/File:US_copper_mine_locations_2003.svg*

Lab 15

Figure L15.1: NOAA, Public domain. *www.lib.noaa.gov/collections/imgdocmaps/daily_weather_maps.html*

Figure in checkout question 2: NOAA, Public domain. *www.lib.noaa.gov/collections/imgdocmaps/daily_weather_maps.html*

Supplementary materials: U.S. Department of Commerce, Daily Weather Maps, Public domain. *www.wpc.ncep.noaa.gov/dailywxmap/index.html*

Lab 16

Figure L16.1: NASA, Public domain. *http://earthobservatory.nasa.gov/IOTD/view.php?id=7205*

Lab 17

Figure L17.1: U.S. Environmental Protection Agency, Wikimedia Commons, Public domain. *https://commons.wikimedia.org/wiki/File:Earth%27s_greenhouse_effect_%28US_EPA,_2012%29.png*

Figure L17.2: PhET Interactive Simulations, Univeristy of Colorado—Bolder, *https://phet.colorado.edu/en/simulation/legacy/greenhouse*

Lab 18

Figure L18.1: NOAA, Public domain. *www.ncdc.noaa.gov/temp-and-precip/state-temps*

Figure L18.2: NASA's Goddard Space Flight Center/Ludovic Brucker, Wikimedia Commons, Public domain. *https://commons.wikimedia.org/wiki/File:An_ice_core_segment.jpg*

Lab 19

Figure L19.1: Modified from Kaboom88, Wikimedia Commons, Public domain. *https://upload.wikimedia.org/wikipedia/commons/c/ca/Blank_US_map_borders.svg*

Figure in checkout question 3: Modified from AlexCovarrubias, Wikimedia Commons, Public domain. *https://en.wikipedia.org/wiki/File:North_America_second_level_political_division.svg*

Lab 20

Figure L20.1: Jacques Descloitres, MODIS Rapid Response Team, NASA/GSFC, Wikimedia Commons, Public domain. *https://commons.wikimedia.org/wiki/File:Hurricane_Isabel_14_sept_2003_1445Z.jpg*

Figure L20.2: Nilfanion, Wikimedia Commons, Public domain. *https://commons.wikimedia.org/wiki/File:Atlantic_hurricane_tracks_1980-2005.jpg*

Two figures in checkout question 1: (Hurricane David) National Hurricane Center, Wikimedia Commons, Public domain. https://commons.wikimedia.org/wiki/File:David_1979_track.png; (Hurricane Andrew) National Hurricane Center, Wikimedia Commons, Public domain. https://commons.wikimedia.org/wiki/File:Andrew_1992_track.png

Lab 21

Figure L21.1: Ks0stm, Wikimedia Commons, CC BY-SA 3.0, *https://upload.wikimedia.org/wikipedia/commons/9/93/May_20%2C_2013_Moore%2C_Oklahoma_tornado.JPG*

Figure L21.2: Maj. Geoff Legler, Oklahoma National Guard, Wikimedia Commons, Public domain. *https://upload.wikimedia.org/wikipedia/commons/d/d3/Aerial_view_of_2013_Moore_tornado_damage.jpg*

Lab 22

Figure L22.1: U.S. Environmental Protection Agency, Wikimedia Commons, Public domain. *https:// commons.wikimedia.org/wiki/File:Earth%27s_ greenhouse_effect_%28US_EPA,_2012%29.png*

Figure L22.2: Narayanese and NOAA, Wikimedia Commons, CC BY-SA 3.0, *https://commons. wikimedia.org/wiki/File:Mauna_Loa_Carbon_ Dioxide_Apr2013.svg*

Figure L22.3: Climate Interactive, *C-ROADS, www. climateinteractive.org/tools/c-roads*

Lab 23

Figure L23.2: Kuczora, Wikimedia Commons, CC BY-SA 3.0, *https://commons.wikimedia.org/wiki/ File:Hoover_sm.jpg*

Figure L23.3: Hans Hillewaert, Wikimedia Commons, CC BY-SA 3.0, *https://commons.wikimedia.org/wiki/ File:Aquifer_en.svg*

Figure L23.4: Shannon, Wikimedia Commons, GFDL 1.2, *https://commons.wikimedia.org/wiki/ File:Coloradorivermapnew1.jpg*

Figure L23.5: Arizona State University, *WaterSim, https://sustainability.asu.edu/dcdc/watersim*